Rainer Mautz

Indoor Positioning Technologies

Rainer Mautz

Indoor Positioning Technologies

A Survey

Südwestdeutscher Verlag für Hochschulschriften

Impressum / Imprint
Bibliografische Information der Deutschen Nationalbibliothek: Die Deutsche Nationalbibliothek verzeichnet diese Publikation in der Deutschen Nationalbibliografie; detaillierte bibliografische Daten sind im Internet über http://dnb.d-nb.de abrufbar.
Alle in diesem Buch genannten Marken und Produktnamen unterliegen warenzeichen-, marken- oder patentrechtlichem Schutz bzw. sind Warenzeichen oder eingetragene Warenzeichen der jeweiligen Inhaber. Die Wiedergabe von Marken, Produktnamen, Gebrauchsnamen, Handelsnamen, Warenbezeichnungen u.s.w. in diesem Werk berechtigt auch ohne besondere Kennzeichnung nicht zu der Annahme, dass solche Namen im Sinne der Warenzeichen- und Markenschutzgesetzgebung als frei zu betrachten wären und daher von jedermann benutzt werden dürften.

Bibliographic information published by the Deutsche Nationalbibliothek: The Deutsche Nationalbibliothek lists this publication in the Deutsche Nationalbibliografie; detailed bibliographic data are available in the Internet at http://dnb.d-nb.de.
Any brand names and product names mentioned in this book are subject to trademark, brand or patent protection and are trademarks or registered trademarks of their respective holders. The use of brand names, product names, common names, trade names, product descriptions etc. even without a particular marking in this works is in no way to be construed to mean that such names may be regarded as unrestricted in respect of trademark and brand protection legislation and could thus be used by anyone.

Coverbild / Cover image: www.ingimage.com

Verlag / Publisher:
Südwestdeutscher Verlag für Hochschulschriften
ist ein Imprint der / is a trademark of
AV Akademikerverlag GmbH & Co. KG
Heinrich-Böcking-Str. 6-8, 66121 Saarbrücken, Deutschland / Germany
. Email: info@svh-verlag.de

Herstellung: siehe letzte Seite /
Printed at: see last page
ISBN: 978-3-8381-3537-3

Zugl. / Approved by: Zürich, ETH, Habilitation, 2012

Copyright © 2012 AV Akademikerverlag GmbH & Co. KG
Alle Rechte vorbehalten. / All rights reserved. Saarbrücken 2012

Abstract

In the age of automation the ability to navigate persons and devices in indoor environments has become increasingly important for a rising number of applications. With the emergence of global satellite positioning systems, the performance of outdoor positioning has become excellent, but many mass market applications require seamless positioning capabilities in all environments. Therefore indoor positioning has become a focus of research and development during the past decade.

It has by now become apparent that there is no overall solution based on a single technology, such as that provided outdoors by satellite-based navigation. We are still far away from achieving cheap provision of global indoor positioning with an accuracy of 1 meter. Current systems require dedicated local infrastructure and customized mobile units. As a result, the requirements for every application must be analyzed separately to provide an individually tailored solution. Therefore it is important to assess the performance parameters of all technologies capable of indoor positioning and match them with the user requirements which have to be described precisely for each application. Such descriptions must be based on a market analysis where the requirements parameters need to be carefully weighed against each other. The number of relevant requirements parameters is large (e.g. accuracy, coverage, integrity, availability, update rate, latency, costs, infrastructure, privacy, approval, robustness, intrusiveness etc.). But also the diversity of different technologies is large, making it a complex process to match a suitable technology with an application. At the highest level, all technologies can be divided into categories employing three different physical principles: inertial navigation (accelerometers and gyroscopes maintaining angular momentum), mechanical waves (i.e. audible and ultra-sound) and electromagnetic waves (i.e. using the visible, infrared, microwave and radio spectrum). Systems making use of the radio spectrum include FM radios, radars, cellular networks, DECT phones, WLAN, ZigBee, RFID, ultra-wideband, high sensitive GNSS and pseudolite systems.

This thesis categorizes all sighted indoor positioning approaches into 13 distinct technologies and describes the measuring principles of each. Individual approaches are characterized and key performance parameters are quantified. For a better overview, these parameters are briefly compared in table form for each technology.

Contents

1 **Introduction** 1
    1.1   Motivation 1
    1.2   Previous Surveys 2
    1.3   Overview of Technologies 3
    1.4   Indoor Positioning Applications 4
    1.5   Structure of this Work 8

2 **User Requirements** 10
    2.1   Requirements Parameters Overview 10
    2.2   Positioning Requirements Parameters Definition 12
    2.3   Man Machine Interface Requirements 15
    2.4   Security and Privacy Requirements 16
    2.5   Costs 16
    2.6   Generic Derivation of User Requirements 16
    2.7   Requirements for Selected Indoor Applications 17

3 **Definition of Terms** 21
    3.1   Disambiguation of Terms for Positioning 21
    3.2   Definition of Technical Terms 23
    3.3   The Basic Measuring Principles 26
    3.4   Positioning Methods 28

4 **Cameras** 31
    4.1   Reference from 3D Building Models 32
    4.2   Reference from Images 33
    4.3   Reference from Deployed Coded Targets 34
    4.4   Reference from Projected Targets 36
    4.5   Systems without Reference 37
    4.6   Reference from Other Sensors 38
    4.7   Summary on Camera Based Indoor Positioning Systems 38

5 **Infrared** 40
    5.1   Active Beacons 40
    5.2   Imaging of Natural Infrared Radiation 41

	5.3	Imaging of Artificial Infrared Light	42
	5.4	Summary on Infrared Indoor Positioning Systems	42
6		**Tactile and Combined Polar Systems**	**43**
	6.1	Tactile Systems	43
	6.2	Combined Polar Systems	44
	6.3	Summary on Tactile and Combined Polar Systems	47
7		**Sound**	**48**
	7.1	Ultrasound	48
	7.2	Audible Sound	54
	7.3	Summary on Sound Systems	54
8		**WLAN / Wi-Fi**	**56**
	8.1	Propagation Modeling	56
	8.2	Cell of Origin	57
	8.3	Empirical Fingerprinting	57
	8.4	WLAN Distance Based Methods (Pathloss-Based Positioning)	60
	8.5	Summary on WLAN Systems	64
9		**Radio Frequency Identification**	**65**
	9.1	Active RFID	66
	9.2	Passive RFID	66
	9.3	Summary on RFID Systems	68
10		**Ultra-Wideband**	**69**
	10.1	Range Estimation Using UWB	70
	10.2	Multipath Mitigation Using UWB	71
	10.3	Positioning Methods Using UWB	72
	10.4	Commercial UWB Systems	74
	10.5	Summary on Ultra-Wideband Systems	74
11		**High Sensitive GNSS / Assisted GNSS**	**76**
	11.1	Signal Attenuation	76
	11.2	Assisted GNSS	77
	11.3	Long Integration and Parallel Correlation	78
	11.4	Summary on High Sensitive GNSS	79
12		**Pseudolites**	**81**
	12.1	Pseudolites Using Signals Different to GNSS	82

| 12.2 | GNSS Repeaters | 83 |
| 12.3 | Summary on Pseudolite Systems | 84 |

13 Other Radio Frequency Technologies 86

13.1	ZigBee	86
13.2	Bluetooth	87
13.3	DECT Phones	88
13.4	Digital Television	88
13.5	Cellular Networks	89
13.6	Radar	91
13.7	FM Radio	94
13.8	Summary on Radio Systems	95

14 Inertial Navigation Systems 96

14.1	INS Navigation without External Infrastructure	96
14.2	Pedestrian Dead Reckoning	97
14.3	INS Pedestrian Navigation Using Complementary Sensors	99
14.4	Foot Mounted Pedestrian Navigation	102
14.5	Summary on INS Based Systems	104

15 Magnetic Localization 105

15.1	Systems Using the Antenna Near Field	105
15.2	Systems Using Magnetic Fields from Currents	106
15.3	Systems Using Permanent Magnets	108
15.4	Systems Using Magnetic Fingerprinting	108
15.5	Summary on Magnetic Localization	109

16 Infrastructure Systems 110

16.1	Power Lines	110
16.2	Floor Tiles	111
16.3	Fluorescent Lamps	111
16.4	Leaky Feeder Cables	112
16.5	Summary on Infrastructure Systems	112

17 Concluding Remarks 114

| 17.1 | Conclusion | 114 |
| 17.2 | Outlook | 114 |

Acronyms 115

Symbols 118

References 119

1 Introduction

Subsequent to the 2010 and 2011 International Conferences on Indoor Positioning and Indoor Navigation (IPIN), the author was repeatedly asked to provide keynote presentations to give an overview of current indoor positioning technologies. An obvious lack of available information on this topic inspired the idea to create this survey of existing techniques for indoor positioning and navigation. An attempt is being made to comprehensively describe relevant approaches, developments and products, at the expense of omitting technical details. Cited references provide such details for each specific system approach. To guide the reader in the process of selecting an appropriate technology, the system parameters and typical performance levels are compared to each other.

Systems based on micro- and nanomeasuring technologies for applications with measuring ranges below 1 m have not been included in this survey. The reason is that developments of small-scale technologies are mainly driven by the manufacturers' research departments and therefore remain unpublished solutions.

An extensive list of application areas is given in Section 1.4. It reveals the significance of indoor positioning to our society and explains the necessity for further research efforts to put these applications into practice.

1.1 Motivation

Following the achievements of satellite-based location services in outdoor applications the challenge has shifted to the provision of such services for the indoor environment. However, the ability to locate objects and people indoors remains a substantial challenge, forming the major bottleneck preventing seamless positioning in all environments. Many indoor positioning applications are waiting for a satisfactory technical solution. Improvements in indoor positioning performance have the potential to create unprecedented opportunities for businesses.

The question why this work draws a distinction between indoor and outdoor positioning has been raised. In fact, most positioning systems can – at least theoretically – be used indoors as well as outdoors. However system performances differ greatly, because the environments have a number of substantial dissimilarities. Indoor environments are particularly challenging for positioning, i.e. position finding, for several reasons:

- severe multipath from signal reflection from walls and furniture
- Non-Line-of-Sight (NLoS) conditions
- high attenuation and signal scattering due to greater density of obstacles
- fast temporal changes due to the presence of people and opening of doors
- high demand for precision and accuracy

On the other hand, indoor settings facilitate positioning and navigation in many ways:

- small coverage areas
- low weather influences such as small temperature gradients and slow air circulation
- fixed geometric constraints from planar surfaces and orthogonality of walls
- infrastructure such as electricity, internet access, walls suitable for target mounting
- lower dynamics due to slower walking and driving speeds.

Another reason why indoor positioning has increasingly become a focus of research is that the dominating technologies for positioning in outdoor environments, namely GNSS (Global Navigation Satellite Systems), perform poorly within buildings. The indoor environment lacks a system that possesses the excellent performance parameters of outdoor GNSS in terms of global coverage, high accuracy, short latency, high availability, high integrity and low user-costs. Like indoor settings, certain outdoor environments are not well covered by GNSS due to insufficient views to the open sky. Therefore, positioning systems targeting 'GNSS challenged' outdoor environments have been included in this study. Precisely speaking, this survey aims to describe all positioning techniques relevant to challenging environments – even including GNSS approaches suitable for such environments. For simplicity however, the term *indoor positioning* is kept throughout this report.

1.2 Previous Surveys

Hightower and Borriello (2001) set up a classification scheme in order to help developers of location-aware applications to better evaluate their options when choosing a location-sensing system. At this early stage in the development of indoor positioning systems, 15 systems were compared in terms of accuracy, precision, scale, costs and limitations. The quantifications given 10 years ago are hardly valid today. The rapid progress in this emerging field requires a new survey every 3 to 5 years in order to represent a useful state-of-the-art guide.

An extensive survey of wireless indoor positioning techniques and solutions has been carried out by Liu et al. (2007). Their survey details the state-of-the-art in 2005 of GPS, RFID, Cellular-Based, UWB, WLAN and Bluetooth technologies. The performance parameters of 20 systems and solutions are compared in terms of accuracy, precision, complexity, scalability and robustness.

The textbook of Bensky (2007) describes radio-navigation techniques comprehensively and provides details on methods for distance estimation between radios.

A survey of the mathematical methods used for indoor positioning can be found in Seco et al. (2009). The study focuses on wireless positioning techniques grouped into the four categories: geometry-based methods, cost-function minimization, fingerprinting and Bayesian techniques.

Mautz (2009) evaluated 13 different indoor positioning solutions with focus on high precision technologies operating in the mm to cm level. The evaluation is carried out from the perspective of a geodesist and includes the criteria accuracy, range, signal frequency, principle, market maturity and acquisition costs.

These surveys demonstrate conceptual heterogenity, differences in market maturity, variety in the application addressed and dissimilarities in design. Therefore it is difficult – if not impossible – to accomplish objective performance benchmarking.

1.3 Overview of Technologies

All system approaches described in this work have been divided into 13 different technologies. Accordingly, each chapter is dedicated to a distinctive indoor positioning technology. Even if the technology employed is of minor importance to the user, the choice for this categorization is that systems using the same technology can be easily compared in their performance parameters.

Table 1.1 characterizes the sensor technologies at high-level. The values specified for accuracy and coverage are given in form of intervals wherein most approaches reside. There are many exceptions exceeding these intervals. Similarly, only the main measuring principles and applications are mentioned in the table. More details can be found in the tables found in the individual chapters.

Table 1.1 Overview of indoor positioning technologies. Coverage refers to ranges of single nodes.

Chapter / Technology	Typical Accuracy	Typical Coverage (m)	Typical Measuring Principle	Typical Application	Page
4 Cameras	0.1mm – dm	1 – 10	angle measurements from images	metrology, robot navigation	31
5 Infrared	cm – m	1 – 5	thermal imaging, active beacons	people detection, tracking	40
6 Tactile & Polar Systems	µm – mm	3 – 2000	mechanical, interferometry	automotive, metrology	43
7 Sound	cm	2 – 10	distances from time of arrival	hospitals, tracking	48
8 WLAN / WiFi	m	20 – 50	fingerprinting	pedestrian navigation, LBS	56
9 RFID	dm – m	1 – 50	proximity detection, fingerprinting	pedestrian navigation	65
10 Ultra-Wideband	cm – m	1 – 50	body reflection, time of arrival	robotics, automation	69
11 High Sensitive GNSS	10 m	'global'	parallel correlation, assistant GPS	location based services	76
12 Pseudolites	cm – dm	10 – 1000	carrier phase ranging	GNSS challenged pit mines	81
13 Other Radio Frequencies	m	10 – 1000	fingerprinting, proximity	person tracking	86
14 Inertial Navigation	1 %	10 – 100	dead reckoning	pedestrian navigation	96
15 Magnetic Systems	mm – cm	1 – 20	fingerprinting and ranging	hospitals, mines	105
16 Infrastructure Systems	cm – m	building	fingerprinting, capacitance	ambient assisted living	110

1 Introduction

A graphical overview in dependence of accuracy and coverage is given in Figure 1.1. The coverage is to be regarded as the direct measuring range of an unextended implementation, i.e. the spatial scalability which many system approaches offer has not been taken into account (e.g. deployment of additional sensor nodes). If a system architecture includes a combination of different sensor technologies (e.g. inertial navigation and WLAN), then the work is described under the chapter with the technology that is most significant to the system approach.

Most technologies rely on electromagnetic waves and a few on mechanical (sound) waves. As can be seen from Figure 1.2 a large part of the electromagnetic spectrum can be exploited for indoor positioning. High accuracy systems tend to employ shorter wavelengths.

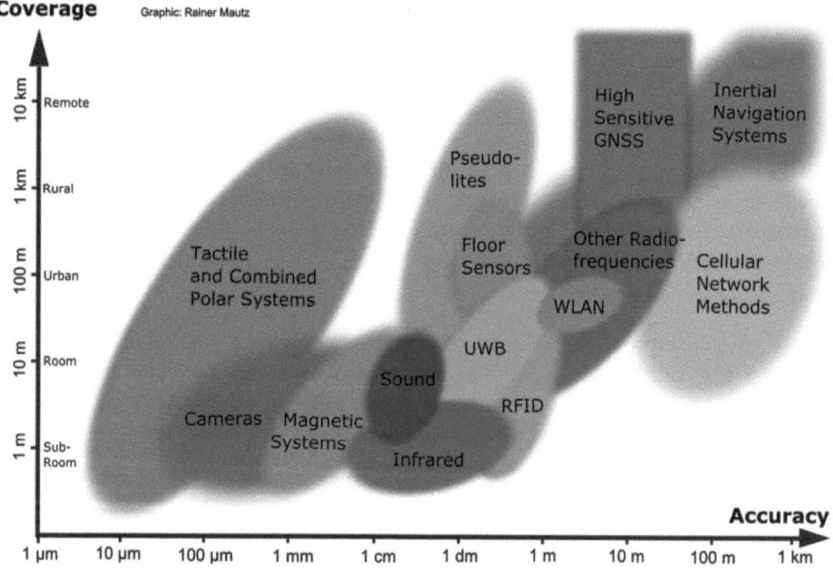

Figure 1.1 Overview of indoor technologies in dependence on accuracy and coverage

1.4 Indoor Positioning Applications

The list of applications below demonstrates the omnipresent need for indoor positioning capability in our modern way of life. Moreover, along with an improvement of performance, future generations of indoor positioning systems will find even more applications which are at the present time not feasible.

1.4.1 Location Based Services in Indoor Environments

Commercially highly relevant applications for the mass market are the so-called Location-Based Services (LBS) which make use of the geographical position to deliver context-dependent information accessible with a mobile device. Such services are required indoors and outdoors. Examples of indoor LBS are obtaining safety information or topical information on cinemas, concerts or events in the vicinity. LBS applications include navigation to the right store in a mall or office in a public building. Within a store or warehouse, the location detection of products is of interest to the owner as well as to the customers. In particular, location-based advertisements, location-based billing and local search services have a high commercial value. At large tradeshows, there is a request to guide the visitors to the correct exposition booths. Applications at train or bus stations include the navigation to the right platform or bus stop. Further examples of LBS are proximity-based notification, profile matching and the implementation of automated logon/logoff procedures in companies. There is also added value for the positioning provider, e.g. by resource tracking, fleet management and user statistics.

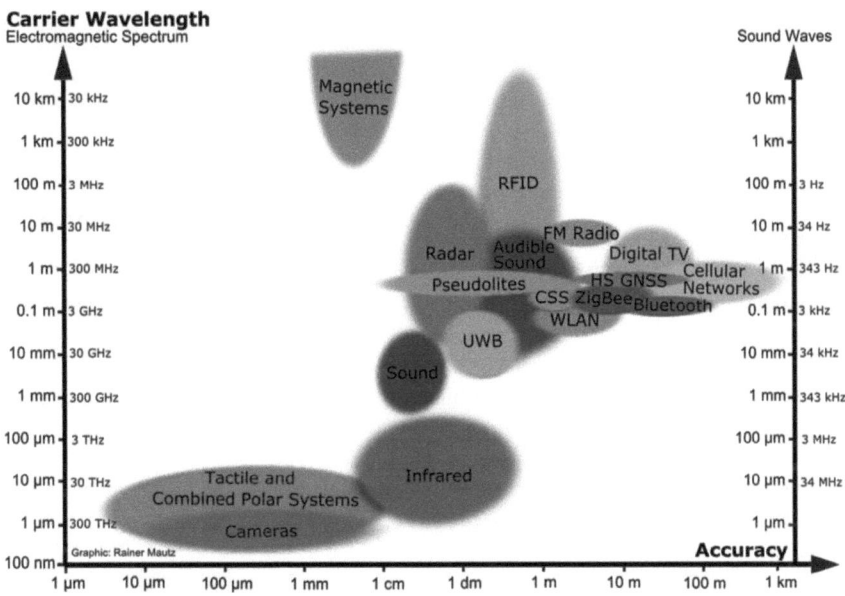

Figure 1.2 Indoor technologies in dependence on accuracy and carrier wavelength

1.4.2 Private Homes

Applications at homes include the detection of lost items, physical gesture games and location based services at home. Ambient Assistant Living (AAL) systems provide assistance for elderly people in their homes within their activities of daily living. A key function of AAL systems is location awareness which requires an indoor positioning functionality. Applications at home are medical monitoring such as monitoring vital signs, detection of emergencies and fall detection, but also service and personalized entertainment systems, such as smart audio systems (Zetik et al. 2010).

1.4.3 Context Detection and Situational Awareness

Mobile devices provide a large variety of useful functions where it is desirable to have an automated adaptation of the mobile device depending on a change of the user's context. Such functionality spares the user additional effort by providing assistance in individual situations. To enable such an automatic adaptation the mobile user's context needs to be determined by the mobile device itself. The most significant criteria to determine the user's context is the current geographical location. For example a smart conference guide can provide information about the topic discussed in nearby auditoriums.

1.4.4 Medical Care
In hospitals the location tracking of medical personnel in emergency situations has become increasingly important. Medical applications in hospital also include patient and equipment tracking, e.g. fall detection of patients. Precise positioning is required for robotic assistance during surgeries. Existing analytical devices can be replaced with more efficient surgical equipment.

1.4.5 Social Networking
As a member of the young generation participation in the network has become increasingly important because social integration is governed through the social network. Ubiquitous location plays a central role in social networking, such as locating friends for coordinating joint activities.

1.4.6 Environmental Monitoring
Environmental monitoring is used to observe some phenomenon such as heat, pressure, humidity, air pollution and deformation of objects and structures. To monitor these parameters over a certain indoor or outdoor space, multiple sensor nodes are organized as a Wireless Sensor Network (WSN). A WSN consists of small, inexpensive, spatially distributed autonomous nodes with limited processing and computing resources and radios for wireless communication. A comprehensive literature review on WSNs can be found in Yick et al. (2008). In order to retrieve the nodes' positions from ranging and proximity information among these sensor nodes, dedicated algorithms of cooperative localization have been developed, see Mautz et al. (2007a).

1.4.7 Police and Firefighters
Indoor positioning capabilities provide important benefits in law enforcement, rescue services, and fire services i.e. location detection of firemen in a building on fire. The police benefits from several relevant applications, such as instantaneous detection of theft or burglary, detection of the location of police dogs trained to find explosives in a building, locating and recovery of stolen products for post-incident investigations, crime scene recovery, statistics and training but also in the prevention of crime, e.g. with tagged devices for establishing so-called geofenceing i.e. alarm systems which can detect whether a person or an asset has left a certain area unauthorized.

1.4.8 Intelligent Transportation
A mass user application for vehicles will be the provision of seamless navigation through extension of road guidance inside parking garages (Wagner et al. 2010). In particular, it becomes possible to navigate the driver to a single parking spot and from there to the pedestrian destination (Gusenbauer et al. 2010).

1.4.9 Industry

Mechanical engineering is developing towards intelligent systems for more or less fully automatic manufacturing. For numerous industrial applications indoor position awareness is an essential functional element, such as for robotic guidance, industrial robots, robot cooperation, smart factories (e.g. tool assistance systems at car assembly lines), automated monitoring and quality control. Indoor positioning capabilities can help to find tagged maintenance tools and equipment scattered all over a plant in industrial production facilities. The improvement of automatic safety systems, intelligent worker protection and collision avoidance is driven by the positioning capability of such a system.

1.4.10 Museums

There are several applications in museums, such as visitor tracking for surveillance and study of visitor behavior, location based user guiding and triggered context aware information services.

1.4.11 Financial Institutions

For the seamless documentation of valuables during their transport, an indoor tracking component is required.

1.4.12 Logistics and Optimization

For the purpose of process optimization in complex systems, it is essential to have information about the location of assets and staff members. In a complex storage environment for example, it is important that requested goods are found quickly. Based on accurate localization, tracing of every single unit becomes possible. Positioning for cargo management systems at airports, ports and for rail traffic affords unprecedented opportunities for increasing their efficiency.

1.4.13 Guiding of the Vulnerable People

Systems designed specifically to aid the visually impaired should operate seamlessly in all indoor and outdoor environments. Navigation is generally required for vulnerable people to assist walking in combination with public transport.

1.4.14 Structural Health Monitoring

Sensors incorporated into steel reinforcements within concrete can perform strain measurements with high resolution. Strain sensing systems based on passive sensor-integrated RFIDs can measure strain changes and deformation caused by loading and deterioration (OKI 2011).

1.4.15 Surveying and Geodesy

Surveying of the building interior includes setting out and geometry capture of new buildings as well as for reconstructions. Positioning capabilities with global reference are needed for data input to CAD, GIS or CityGML. Accuracy requirements vary from centimeters to millimeters.

1.4.16 Construction Sites

Apart from surveying applications, large constructions sites require positioning capabilities that can support an information management system. The capability to localize and track workers is a crucial component to establish an automatic safety system.

1.4.17 Underground Construction

Special positioning requirements apply in dusty, dark, humid and space limited environments for tunneling (Schneider 2010) and longwall mining (Fink et al. 2010).

1.4.18 Scene Modeling and Mapping

Scene modeling – the task of building digital 3D models of natural scenes – requires the precise orientation of the optical sensor. Indoor mapping systems need to know the camera's position in order to merge multiple views and generate 3D point clouds. Scene modeling is beneficial for several applications such as computer animation, notably virtual training, geometric modeling for physical simulation, mapping of hazardous sites and cultural heritage preservation.

1.4.19 Motion Capturing

Motion capturing relies on the detection of physical gestures and the capability to locate and track body parts. Such technologies are useful for medical studies and animated films. Location based gaming, such as exergaming (gaming as a form of exercise) relies on tracking body movement or reaction of the players.

1.4.20 Applications Based on Augmented Reality

Localization awareness is of fundamental importance for Augmented Reality (AR) applications – an increasingly powerful tool to superimpose graphics or sounds on the users' view, allowing the user to perceive overlaid information which is spatially and semantically related to the environment. An example of vision based navigation for AR is presented in Kim and Jun (2008).

1.4.21 Further Applications

Applications areas which have not been explicitly mentioned above are self-organizing sensor networks, ubiquitous computing, computer vision, industrial metrology, architecture, archeology, civil engineering, pipe inspection (i.e. locating pipes) and facility management.

1.5 Structure of this Work

This introduction is followed by an overview of the user requirement parameters for indoor positioning applications in Chapter 2. The key requirements are defined, a generic method for derivation of requirements is shown and the requirements of some selected applications are quantified. Chapter 3 defines technical terms frequently used in the field of indoor positioning. The basic measuring principles and positioning methods are briefly described. Chapters 4–16 are devoted to a more detailed presentation of different technologies. Each chapter introduces an individual technology and characterizes some representative system implementations. At the end of each chapter a short conclusion summarizes the findings and provides an overview of the key parameters in table form. Chapter 17 closes the thesis with some general conclusions drawn from the presented literature, along with a suggestion on how the current insufficiency in system performances can be systematically improved.

2 User Requirements

A crucial element for any initiative to design an indoor positioning system is a thorough study of the user requirements and specific application descriptions in order to justify the research and development in this field. Requirements for significant applications should drive the future direction of research. Therefore it is important to state well-grounded figures of requirements parameters and allocate suitable technologies.

In this chapter an overview of the user requirement parameters is given in Section 2.1 and a more comprehensive definition of the key requirements can be found in Section 2.2. In addition, a generic method to determine the values for a specific application is indicated. The chapter is concluded in Section 2.7 by summarizing results of different studies on indoor positioning requirements.

2.1 Requirements Parameters Overview

The following list of different parameters can be used as a basis for assessment and comparison of different indoor positioning systems. Due to the large number of criteria, it is not straightforward for a user to identify the optimal system for a particular application. Figure 2.1 illustrates the complexity and multi-dimensionality of the optimization problem confronting the user. For each application, the 16 user requirements need to be weighted against each other. The different requirements are listed below with some example values given in brackets. Apart from these user requirements, there are other important technical parameters of indoor positioning systems such as those shown in Figure 2.2.

In order to serve market needs the embedded technology should be adequately low-cost, low-power, low-latency, miniaturized, require low maintenance and minimal amount of dedicated infrastructure. Research often neglects issues such as security, privacy and reliability.

2.1.1 List of the most Important User Requirements
- accuracy / measurement uncertainty (mm, cm, dm, meter, decameter level)
- coverage area / limitations to certain environments (single room, building, city, global)

2.1 Requirements Parameters Overview

- cost (unique system set-up costs, per user device costs, per room costs, maintenance costs),
- required infrastructure (none, markers, passive tags, active beacons, pre-existing or dedicated, local or global),
- market maturity (concept, development, product)
- output data (2D-, 3D coordinates, relative, absolute or symbolic position, dynamic parameters such as speed, heading, uncertainty, variances)
- privacy (active or passive devices, mobile or server based computation)
- update rate (on-event, on request or periodically e.g. 100 Hz or once a week)
- interface (man-machine interfaces such as text based, graphical display, audio voice and electrical interfaces such as RS-232, USB, fiber channels or wireless communications)
- system integrity (operability according technical specification, alarm in case of malfunction)
- robustness (physical damage, theft, jamming, unauthorized access)
- availability (likelihood and maximum duration of outages)
- scalability (not scalable, scalable with area-proportional node deployment, scalable with accuracy loss),
- number of users (single user e.g. totalstation, unlimited users e.g. passive mobile sensors),
- intrusiveness / user acceptance (disturbing, imperceptible)
- approval (legal system operation, certification of authorities)

Figure 2.1 User requirements

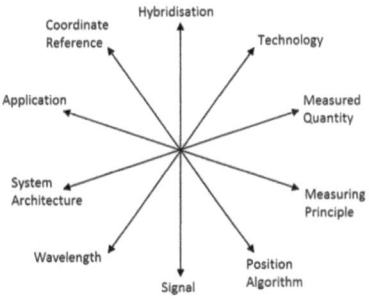

Figure 2.2 Important technical parameters being less significant to the user

2.1.2 Technical Parameters Less Important to the User

- level of hybridization (single modality, two different sensors, highly hybrid sensor fusion).
- technology (optical, inertial, magnetic, sound, ...)
- measured quantity (direction, distance, signal amplitude, acceleration, time)

2 User Requirements

- basic measuring principle (tri-)lateration, (tri-)angulation, fingerprinting, cell of origin, dead-reckoning)
- positioning algorithm used (multidimensional scaling, multilateration, heuristics)
- signal used (sound waves, electromagnetic waves, magnetic field strength)
- signal wavelength (visible light, infrared, radio frequencies)
- system architecture (central or distributed systems)
- application (navigation, surveying, industry tracking, metrology)
- coordinate reference (local, global, object or sensor coordinate system)

2.1.3 Evaluation of Positioning Systems

In order to find a suitable positioning technology for a particular application, the performance parameters need to be matched with the user requirements. These parameters (listed above and detailed in Section 2.2) pose a multidimensional optimization problem when searching for the best match. Moreover, the values for the performance parameters are usually not exactly determinable since they in turn depend on various factors, circumstances and conditions. Each system approach has not only its individual set of performance parameters, but also several unique characteristics, conditions, assumptions and applications which need to be weighted against each other. Weighting of all parameters and additional conditions cannot be done in an objective manner. Therefore, fair-minded ranking of the systems is neither useful nor feasible.

2.2 Positioning Requirements Parameters Definition

2.2.1 Accuracy / Measurement Uncertainty

The accuracy of a system is an important user requirement which should be quantified in any description of an application. The term accuracy has been defined in the Joint Committee for Guides in Metrology (JCGM) as the closeness of agreement between a measured quantity value and a true quantity value of a measurand. In the new concept of measurement uncertainty published in JCGM 200:2008 (2008) the term 'true value' has been discarded. In accordance, 'measurement accuracy' is not used anymore for quantification of a numerical quantity. Instead of 'measurement accuracy' the term 'measurement uncertainty' is used now for quantification of a standard deviation (including the two categories Type A and Type B). Measurement uncertainty comprises, in general, many components. Only some of these components (Type A) may be evaluated from the statistical distribution. Components evaluated from probability density functions based on experience or other information belong to Type B. In order to take into account all components of uncertainty, including those arising from systematic effects, such as components associated with corrections, all systematic measurement errors must be modeled and calibration must be completed by means of a measured quantity value having a negligible measurement uncertainty. However, researchers, developers and vendors still quantify the performance of indoor positioning systems in terms of 'positioning accuracy'. In order to be able to compare the system performances, the conventional definition of 'positioning accuracy' as reported in the sources is used throughout this book. 'Positioning accuracy' should be understood as the degree of

2.2 Positioning Requirements Parameters Definition

conformance of an estimated or measured position at a given time, to the true value, expressed for the vertical and horizontal components at the 95% confidence level. If normal distribution can be assumed, a useful metric for the quality of positions is the computation of the standard deviation (i.e. RMSD, Root Mean Square Deviation)

$$\sigma_\mathbf{P} = \sqrt{\frac{1}{n}\sum_{i=1}^{n}\left(\hat{\mathbf{P}}_i - \mathbf{P}_i\right)^2}, \qquad (2.1)$$

where n is the number of estimated (i.e. measured) position vectors $\hat{\mathbf{P}}_i$ and $\mathbf{P}_i$ the position vector predicted by a model of the localized node i, or, if only one single location is estimated, $\mathbf{P}_i$ is replaced with a single position vector $\mathbf{P}_0$. A criterion which is less sensitive to outliers is the average absolute position deviation

$$a_\mathbf{P} = \frac{1}{n}\sum_{i=1}^{n}\left|\hat{\mathbf{P}}_i - \mathbf{P}_i\right|. \qquad (2.2)$$

In most cases a predicted location $\mathbf{P}_i$ in Equations (2.1) and (2.2) is represented by an empirical mean value. If the unknown coordinates are to be estimated from a redundant set of observations, the average of the estimated mean square positional variances

$$\hat{a}_\mathbf{P} = \frac{1}{n}\sum_{i=1}^{n}\sqrt{q_{xxi} + q_{yyi} + q_{zzi}}, \qquad (2.3)$$

can be computed, where q_{xxi}, q_{yyi}, q_{zzi} are diagonal elements of the variance-covariance matrix $\mathbf{C}_x$ of the estimated parameters as a result of a network adjustment. In this book, 'low accuracy' refers to a standard deviation $\sigma_\mathbf{P} > 10$ m and 'high accuracy' to $\sigma_\mathbf{P} < 1$ cm if no value of accuracy is stated explicitly.

Although the accuracy of an indoor positioning system is the key driver for most applications, it needs to be viewed in context with the other performance parameters described below.

2.2.2 Coverage

Describes the spatial extension where system performance must be guaranteed by a positioning system. One of the following categories should be specified:

a) Local Coverage: a small well-defined, limited area which is not extendable (e.g. a single room or building). For this case, the coverage size is specified (e.g. (m), (m^2) or (m^3)).
b) Scalable Coverage: Systems with the ability to increase the area by adding hardware (e.g. through deployment of additional sensors). In this book, the parameter 'coverage' is set to 'scalable' only if the scalability is not affected by a loss of accuracy.
c) Global Coverage: system performance worldwide or within the desired / specified area. Only GNSS systems and celestial navigation belong to this category.

2.2.3 Integrity

Integrity relates to the confidence which can be placed in the output of a system. Integrity risk is the probability that a malfunction in the system leads to an estimated position that differs from the required position by more than an acceptable amount (the alarm limit) and that the user is not informed within the specified period of time (time-to-alarm). Regulatory bodies have studied and defined integrity performance parameters in some sectors such as civil aviation, however, in other sectors, including those relating to indoor navigation it is more difficult to find quantified integrity parameters. From the application description, this requirement parameter should give an indication whether the devices for integrity parameters are related to Safety of Life (SoL), economic factors, or convenience factors. In academic research papers which describe indoor positioning approaches, the integrity parameter is usually not specified. Therefore this survey does not take integrity into account.

2.2.4 Availability

Availability is the percentage of time during which the positioning service is available for use with the required accuracy and integrity. This may be limited by random factors (failures, communications congestion) as well as by scheduled factors (routine maintenance). Generally one of the following three levels could be specified, although this will depend on the particular application:

a) low availability: < 95%
b) regular availability: > 99%
c) high availability: > 99.9%

To achieve availability, it is assumed that continuity, accuracy and integrity requirements are fulfilled. Application descriptions usually include specification of availability, whereas system developers usually do not specify an availability figure.

2.2.5 Continuity

The continuity is the property of continuous operation of the system over a connected period of time to perform a specific function. The frequencies of acceptable outages should be given. The continuity requirement is usually similar to that of availability.

2.2.6 Update Rate

The update rate is the frequency with which the positions are calculated on the device or at an external processing facility. The following types of measurements rates exist:

a) periodic: regular update, specified in an interval (unit e.g. (Hz))
b) on request   : triggered by the user or by a remote device.
c) on event: measurement update initiated by the local device when a specific event occurs, e.g. when a temperature sensor exceeds a critical threshold.

2.2.7 System Latency

The system latency describes the delay with which the requested information is available to the user. The latency can have the following values:

- real time: Does not tolerate 'perceivable' delays. It is the most demanding latency requirement. It is necessary for navigation and almost all indoor positioning applications.
- sooner the better: Requires the system's best effort.
- sooner the better with an Upper Limit: Requires the system's best effort but the system must be designed to limit the maximum delay to a specified threshold.
- post processing: No specific time of delivery is defined.

2.2.8 Data Output

In addition to times and positions, a number of spatio-temporal data derivatives may be required, many of these can be provided without significantly increasing the data capture or storage requirements. The following derived values are of interest in many applications:

- speed / velocity
- acceleration
- heading / bearing
- predicted position

The requirements specification should explicitly mention if the heading of a mobile object is needed. Some applications require the full spatial orientation, e.g. in form of values for 6 Degrees of Freedom (6 DoF, i.e. 3 coordinate and 3 rotation parameters).

2.3 Man Machine Interface Requirements

The man machine interface requirements describe how position information will be reported and queried at the user device. The following questions need to be answered for an application description.

2.3.1 Information Display – Spatial Data Requirements
- Is a graphical display required?
- Is a scaled map required or is topological correctness sufficient?
- Is additional cartographic/mapped information required?
- What level of detail is required approximately?

2.3.2 Data Query and Analysis Tools
- Is on-request status information about the network or devices needed?
- Is route planning information required?
- Is en-route guidance (visual or audio) needed?
- Are Natural Language Instructions (NLI) required? NLI is a convenient way to provide route information to users, offering rich and flexible means of describing navigational paths.

2.4 Security and Privacy Requirements

The figures about security issues should be given. In addition, several aspects of privacy, such as approval by the user need to be considered.

2.4.1 Requirements for Security and Safety

The security of a system is the extent of protection against some unwanted occurrence such as the invasion of privacy, theft, and the corruption of information or physical damage. The quality or state of being protected from unauthorized access or uncontrolled losses or effects should be given. Safety is a property of a device or process which limits the risk of accident below some specified acceptable level.

2.4.2 Requirements for Privacy and Approval

The level of privacy influences the approval by the user: How comfortable are users with their data (e.g. trajectory) being stored? Do users have legal concerns about their privacy? If so, can private users be motivated to provide personal data?

Approval also includes the requirements for the system to allow certification by authorities. E.g. if there is a need for admissibility in court, the requirements for the system to deliver evidence should be given. Insurance companies should point out their policies concerning approval.

2.5 Costs

The maximum cost of a positioning system is an important user requirement which can be assessed in several ways. Time costs include factors such as the time required for installation and administration. Capital costs include factors such as the price per mobile unit or system infrastructure and the salaries of support personnel. Maintenance costs include expenses required to keep the system functional. Space costs involve the amount of installed infrastructure and the hardware's size. The quantification of the costs should be handled with care due to time-, location-, manufacturer-related dependencies.

2.6 Generic Derivation of User Requirements

Figure 2.3 shows the general approach to define user requirements. First, the potential user groups are defined and listed. Based on the user groups, their associated services are determined. Then the minimum high-level functions that a potential positioning system must fulfill are defined. From these high level functions, a list of parameters to capture the user requirements is derived. The data acquisition (step 5) is carried out from a combination of sources. Primarily a user survey is performed with questionnaires, brainstorming sessions and interviews of industry partners. The evaluation of the questionnaires, interviews and sessions with the user groups is then carried out for each user group and application separately. The result of such a study is the summary of the user requirements parameters in an explicit form.

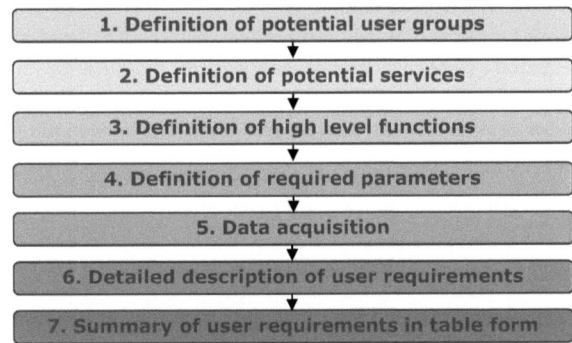

Figure 2.3 Procedure for user requirements capture

2.7 Requirements for Selected Indoor Applications

This section provides numerical targets of some application areas for indoor positioning as stated in various studies of experts. These numbers demonstrate large dissimilarity of user requirements between different applications. Figure 2.4 shows an overview of required accuracies and ranges allowing for direct comparison with the performances of technologies in Figure 1.1 on page 4.

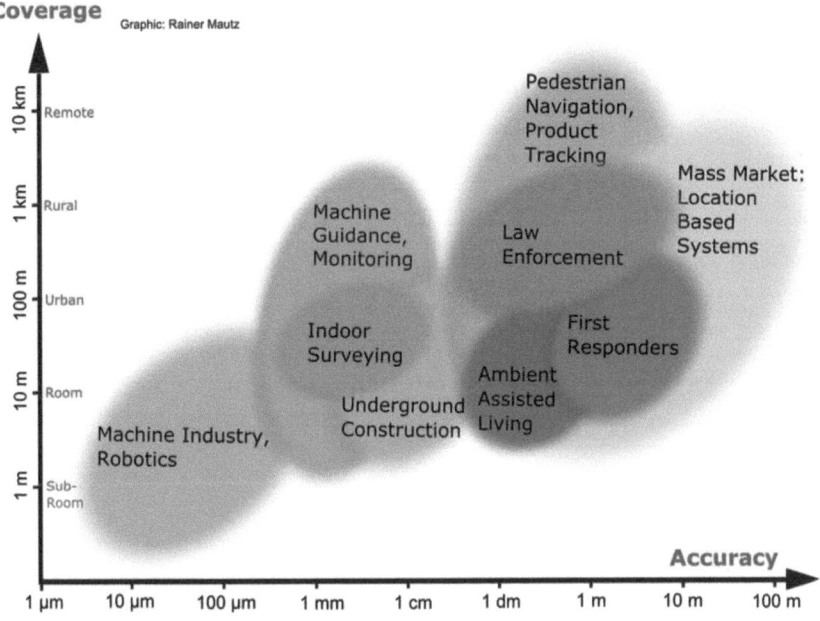

Figure 2.4 Overview of user requirements in terms of accuracy and coverage

2 User Requirements

2.7.1 Requirements for the Mass Market

Mass market applications for indoor positioning require the use of standard devices without supplementary physical components, e.g. major modifications to mobile phones in order to include a positioning function are out-of-scope in the mass market. The general user requirements for mass-market localization have been put into numbers by Wirola et al. (2010), see Table 2.1.

Table 2.1 Summary of requirements for mass-marked localization according to Wirola et al. (2010)

Criteria	Criteria Description	Value
horizontal accuracy	2D position for the detection of a shelf in a supermarket	1 m
vertical accuracy	selection of the correct floor and visualization	floor detection
update rate	minimum for navigation	1 Hz
latency	delay with which position is available to the user	none
TTFF	Time-To-First-Fix, latency after switching on the device	without delay
privacy	maintenance of the user privacy	according to user-set policy

2.7.2 Requirements for Underground Construction

Schneider (2010) details the positioning requirements for underground construction. In contrast to pedestrian navigation applications, the positioning requirements for underground surveying are more demanding in terms of accuracy which needs to be in the order of millimeters instead of meters. Other requirements such as constraints on costs, size and electrical power are therefore less demanding. Additional requirements apply in terms of system robustness. Table 2.2 quantifies the requirements as stated by Schneider.

Table 2.2 Summary of positioning requirements in underground construction according to Schneider (2010)

Criteria	Criteria Description	Value
accuracy	for deformation analysis	1 mm – 5 mm
accuracy	for heading and machine guidance	1 cm – 5 cm
range	depends on the application	20 m – 50 m
3D-positioning	tasks require 3D-coordinates	yes
resistance against perturbation, robustness	required against external impacts such as dust (especially close to tunnel face), emissions from construction machines, damage caused by ongoing construction (e.g. drill & blast), vibrations and tunnel deformations	yes
real-time availability	construction surveying tasks need results in real-time	80%
user friendliness	system should be operable by foremen without surveying background	yes
costs	system cost must not exceed that of a surveying totalstation	10'000 € - 50'000 €
operability under non-line of sight	system must be operable under NLoS conditions, continuous and direct LoS between the reference sensors and the work site is not always given	required
power supply	availability for external electrical power	guaranteed

2.7.3 Requirements for Indoor Surveying

Carpenters, architects, interior designers and fitters would benefit from a tool capable of delivering 3D positions within mm-accuracy. Such a tool must be user-friendly in the sense that the system set-up is quick and wireless operation of a handheld device is possible. Real-time tracking with 20 Hz or more is necessary to allow for capturing profiles and maintaining robustness during fast pivoting movements of the operator. An overview of the user requirements is given in Table 2.3.

Table 2.3 Summary of requirements in indoor surveying applications

Criteria	Criteria Description	Value
accuracy	3D position compared to reference	2 mm (at 20 m)
coverage	3D measurement volume	20 m
size	size of mobile measurement unit	handheld
update rate	high rates needed for tracking and fast movements	20 Hz
operating time	time of battery life	10 hours
installation complexity	time to set-up system	< 2 min
quality indicator	self-reporting of current accuracy	yes
costs	user price per unit	< 3000 €

2.7.4 Requirements for Ambient Assisted Living

Prior to an evaluation of positioning systems for Ambient Assisted Living (AAL) through competitive benchmarking (EvAAL 2011) the user requirements for AAL applications were defined in an open discussion. It revealed a 2D accuracy of 0.5 m to 1 m and an update rate of 0.5 s. An important requirement is the 'user acceptance', which describes how intrusive a system is to the user, e.g. does an elderly person notice the system by wearing tags on the body? An overview of the requirements including their relative weights is given in Table 2.4.

Table 2.4 Summary of requirements in AAL applications

Criteria	Criteria Description	Value	Weight
accuracy	2D position compared to reference	0.5 m – 1 m	0.25
installation complexity	man-minutes to install an AAL system in a flat	< 1 hour	0.20
user acceptance	qualitative measure describing invasiveness	non-invasive	0.20
availability	fraction of time a system is active and responsive	> 90 %	0.15
integrability of AAL	use of standards and open protocols	-	0.10
update rate	Sample interval of the location system	0.5 s	-
coverage	area of a typical flat	90 m^2	-
costs	not assessed within the evaluation	-	-

2.7.5 Requirements for First Responders

Rantakokko et al. (2010) quantify the requirements for enforcement officers, firefighters and military personnel. They identify the following key requirements as stated in Table 2.5.

Table 2.5 Summary of requirements for first responders according to Rantakokko et al. (2010)

Criteria	Criteria Description	Value
horizontal accuracy	need for specific room determination	≤ 1 m
vertical accuracy	need for determination of a specific floor in a building	≤ 2 m
update rate	updates need to provide constant accessibility	permanent
latency	delay with which position is available to the user	none
weight	weight of personal localization and tracking gear	< 1 kg
cost	price of complete positioning system	< €1000

2 User Requirements

Further requirements include physical robustness, encrypted communication, estimation of uncertainty, compatibility with other information sources, real-time map-building, and user friendliness.

2.7.6 Requirements for Law Enforcement

The study of Mautz (2005) describes user requirements for a proposed positioning system in all environments for crime prevention, crime detection and the detection of stolen goods. Table 2.6 gives an overview of the potential services which have been identified and quantifies the required parameters for individual services. Generally speaking, a positioning accuracy of 1 m or better is required indoors for most services.

Table 2.6 Summary of positioning requirements in crime reduction management according to Mautz (2005)

Service	Required Data	Positioning Accuracy (95%) (indoor)	Integrity			Availability (maximum allowable continuous outage)	Update Rate	Brief Justification
			Time to Alarm	Horizontal Alarm Limit				
target hardening, reducing likelihood of a device being stolen	position, ranging, alarm	1 m	10 s	2 m – 20 m		> 95 % (5 min)	on event, in event: 1 s – 10 s	geofencing, alarm leaving designated area or network
reduction in the value of goods.	alarm	--	30 s	--		> 95 % (1 h)	on event, on request	disabling of devices, electronic marker
increasing the risk for criminals of being caught, surveillance	position, ranging, movement, alarm	0.5 m	1 s	2 m – 20 m		> 95 % (1 min)	on event, in event: 1 s – 10 s	motion detection in offices, surveillance on roads and crime hotspots.
instantaneous detection of theft or burglary	ranging, position, speed, heading, track	0.1 m	1 s	1 m – 2 m		> 99 % (1 min)	on event, in event: 1 s – 10 s	movement detection of devices in offices, real-time tracking on streets and roads
locating and recovery of stolen products.	ranging, position, track	0.5 m	60 s	5 m – 10 m		> 99 % (5min)	on event, on request	trajectory of movements, locate stolen products
investigation on crime, e.g. location from wireless digital devices	position, track	5 m	60 s	10 m – 20 m		> 90 % (5 min)	on request	crime scene recovery. locate mobile phones on roads
training in crime detection	position, time	5 m	--	--		> 90 % (60 min)	on request	identification of crime hotspots

3 Definition of Terms

In this chapter prevalent technical terms for 'positioning' are defined and an explanation is given for technical terms frequently used in the field of indoor positioning. The chapter concludes with the definition of basic measuring principles and positioning methods.

3.1 Disambiguation of Terms for Positioning

In the literature, the process of determining a location is described by a number of different technical terms with slightly different meanings. The following definitions primarily reflect their usage in this work and might be defined slightly differently elsewhere.

3.1.1 Positioning

Positioning is the general term for determination of a position of an object or a person. It is particularly used to emphasize that the target object has been moved to a new location.

3.1.2 Localization

Mostly used for describing the process of position determination in wireless sensor networks based on communicating nodes. In the literature, the use of 'localization' to mean 'positioning' emphasizes the fact that positioning is carried out in an ad-hoc and cooperative manner. The term 'localization' also underlines that the application requires topological correctness of the sensor locations, whereas the absolute coordinate position is of minor importance. Therefore, localization is mainly associated with rough estimation of location for low-accuracy systems such as for locating mobile phones.

3.1.3 Wireless Positioning

In the literature, the term 'wireless positioning' refers to radio-navigation techniques that rely on distance estimation. Because all positioning technologies presented here are 'wireless' in the sense that they do not require electric wiring, the term 'wireless positioning' is avoided.

3 Definition of Terms

3.1.4 Geolocation

Geolocation is used for locating internet-connected devices where the determined location is descriptive or context based rather than a set of geographic coordinates.

3.1.5 Location Sensing

As with Geolocation, the term 'Location Sensing' is used in computer science to express information about the location of devices (mostly descriptive, rarely coordinate based) by employing the internet.

3.1.6 Radiolocation

Radiolocation refers to position determination of an object by exploiting intrinsic characteristics of received radio waves. Signal amplitudes, phases, angles of arrival and times of arrival can be used to derive distances.

3.1.7 Locating

Locating is the process of determining the current location of an object or a person relative to a reference position. It is mostly used in conjunction with less accurate location techniques to place, assign or to discover the location of an object.

3.1.8 Positioning

Positioning is a synonym for position finding.

3.1.9 Local Positioning System (LPS) / Real Time Locating System (RTLS) / Active System

The terms LPS, RTLS and 'Active System' refer to beacon-based positioning that depends on locally deployed infrastructure – in contrast to passive systems that use only self-contained sensor data and can operate autonomously. Passive indoor positioning systems normally use inertial sensors in combination with barometers, odometers and magnetometers. Although not operating autonomously, GNSS based navigation is considered as 'passive' since local infrastructure is not required (unless terrestrial differential GNSS is used).

3.1.10 Navigation

Navigation comprises 1) determination of position, speed and heading of a subject or object, 2) finding of the optimal path (in the sense of the fastest, shortest, or cheapest route) from a start to an end location, and 3) guidance along a given path and control of the difference between the current position and the planned path.

Navigation according to definition 2) and 3) requires geoinformation about the navigable indoor space. The two main types of location information are physical and symbolic location. Physical locations are expressed in form of coordinates, which identify a point on a map. A navigation system providing a physical position can usually be augmented to provide corresponding symbolic location information with additional information or infrastructure. Symbolic navigation is based natural-language, such as 'in the library' or 'near the exit door'.

In contrast to tracking, navigation requires position information to be available at the mobile station.

3.1.11 Tracking

The process of repeated positioning of a moving object or person over time is called tracking. In case of object tracking (also denoted as target tracking, path tracking, location tracking, mobile tracking, device tracking or asset tracking) the mobile object is associated with any kind of positioning system. In contrast to navigation, tracking is used when the infrastructure is determining the location of a passive mobile device, where the information about the current position is not necessarily known at the mobile device.

An important special case is that of video tracking (also the terms visual-, optical-, photogrammetric, fiducial marker- or image feature tracking are used). Video tracking involves digital cameras and recognition of target objects in consecutive video frames. High sampling rates of tracking kinematic objects facilitate positioning due to high correlation between consecutive measurements.

The term tracking is often used in conjunction with dead reckoning, where the movement is modeled using previously determined positions, speeds and directions with the aim not only to estimate the current location but also predict future positions.

3.1.12 Network and Mobile Based Positioning

Network based positioning is defined as a technique, where Base Stations (BS) receive signals coming from a Mobile Station (MS). Position determination is carried out remotely on a server within the network. In contrast, mobile based techniques rely on determination of position being exclusively carried out at the MS using the signals it receives from multiple BS. Since the mobile based positioning approach does not require any forwarding of information to the network, it satisfies privacy issues of mobile users. Radio positioning systems are classified as 'network based' or 'mobile based' in correspondence to 'tracking' and 'navigation' respectively.

3.2 Definition of Technical Terms

This section provides definitions for technical terms significant in the field of indoor positioning.

3.2.1 Absolute and Relative Position

A distinction must be made between absolute and relative positions (locations). Absolute locations refer to a global or large area reference grid with its realization in the form of markers, landmarks or GNSS satellites. Absolute coordinate positions refer to such global or superior reference system. In contrast, relative positions depend on a local frame of reference, e.g. coordinates within a small coverage area are given in delta positions to a local realization of reference.

3 Definition of Terms

3.2.2 Known and Unknown Nodes

In the literature, known positions are denoted as anchor nodes, beacons, fixpoints, Access Points (AP), Base Stations (BS) or reference nodes. Typically, these points are represented physically through markers or devices at permanent positions and are fixed to a certain location with known coordinates. The term 'anchor node' is used in larger networks with multi-hop positioning strategies. 'Beacon' is used to indicate active signal emission. 'Fixpoint' refers to stability in time, 'access point' is the technical term used for WLAN nodes, 'base station' for mobile phone architectures and 'reference node' is primarily used to express availability of absolute position information.

In contrast, unknown nodes are denoted as Mobile Stations (MS), Mobile Terminals (MT) or Blind Nodes (BN) and represent those locations whose coordinates need to be determined by the positioning system. Most positioning system architectures allow blind nodes to be mobile. Typically, blind nodes are attached to the user of the system in the form of tags or devices. The unknown node can be any navigable device or can represent a mounted target on a robot.

3.2.3 Centralized and Distributed Positioning

The concept of a centralized system architecture is that the entire position determination is carried out at a central server where all node locations are stored and provided to an operator. Benefits of centralized architectures are simplicity, data consistence, uniform service to all users and lower expansion costs because the input and output modules contain fewer components. In a distributed system, position determination is carried out onboard of each node based on local observations. The advantages of a distributed architecture are unlimited system scalability and guarantee of the user's privacy.

3.2.4 Line of Sight (LoS) and Non Line of Sight (NLoS)

LoS is present when a signal can travel on the direct straight path from an emitter to a receiver. Positioning techniques relying on LoS are common, e.g. time of arrival distance measurements based on radio signals. Due to occlusions from walls, furniture and people, indoor environments typically induce NLoS propagation, which causes inconsistent time delays at a radio receiver. These delays pose a particular challenge which can only be tackled by few positioning techniques.

3.2.5 Cramér-Rao Lower Bound (CRLB)

CRLB in general denotes the lower limit for the minimal variance of any unbiased estimates of an unknown parameter. In the context of positioning algorithms, this fundamental limit bounds the achievable localization accuracy and therefore serves as a benchmark for positioning algorithms.

3.2.6 Received Signal Strength Indicator (RSSI)

Signal attenuation can be exploited for distance estimation based on RSSI values. RSSI are observed RSS (Received Signal Strength) values averaged over a certain sampling period and usually specified as received power P_R in decibels. Based on the attenuation model

$$P_R \propto P_T \frac{G_T G_R}{4\pi d^p} \tag{3.1}$$

the received signal power or signal strength P_R can be used to estimate the distance d of a person or a mobile object. In the model, P_T is the transmitted power at the emitter, G_T and G_R are the antenna gains of transmitter and receiver and p is the path loss exponent. The path loss factor p characterizes the rate of attenuation with the increase of distance d. In free space $p = 2$ ($P_R \sim d^{-2}$) respective $p > 2$ for environments with NLoS multipath. For indoor environments the path loss exponent typically takes values between 4 and 6. On the other hand, a corridor can act as a waveguide, resulting in path loss with $p < 2$. The free space model does not take into account that antennas are typically set up above the ground. In fact, the ground acts as a reflector and therefore the received power differs from that of free space. A mathematical formulation of such a model – known as open field model – can be found in Bensky (2007).

Theoretically, distances d_i (with $i = 1...n$ and $n>2$) which have been estimated from RSSI values to multiple beacons can be used to determine the receiver position by multilateration. In real-world application however, interference, multipath propagation and presence of obstacles and people leads to a complex spatial distribution of RSSI values, which is unfavorable for the estimation of distances from RSSI. Therefore fingerprinting has become more popular than propagation modeling.

3.2.7 Signal to Noise Ratio (SNR) or (S/N)

Signal to Noise Ratio (SNR) is defined as the ratio of signal power to the power level of background noise. An SNR of more than 1 indicates that the signal is stronger than the noise. As with RSSI, SNR can be used for fingerprinting. Instead of absolute signal strength, SNR patterns of the user device are compared to the database.

3.2.8 Indoor Path Loss Model

The so-called ITU Model for Indoor Attenuation takes into account special properties of radio propagation inside buildings. The model provides a relation between distance d and the total path loss L with

$$L = 20 \log f + p \log d + c(k, f) - 28, \tag{3.2}$$

where f is the radio frequency, c an empirical floor loss penetration factor and k the number of floors between transmitter and receiver. A room inside a building is considered as a closed area limited by walls where the signal is reflected, absorbed or is able to penetrate to a certain extent. Since the prediction of the path loss requires complex modeling, the values for the pass loss coefficient p are determined empirically and differ between $p = 20$ and $p = 30$ depending on the type of indoor space.

3.2.9 Multipath Environment

In a multipath environment different paths (echoes) reach the receiver with different time delays. If differences in paths length are less than the reciprocal of the transmission bandwidth, these paths cannot be resolved as distinct pulses and are observed as the envelope of their sum.

3.2.10 Multipath Modeling

Multipath propagation of signals is particularly problematic for time based ranging methods. In indoor environments where NLoS is the standard case, time of flight estimation requires multipath propagation modeling and mitigation. The problem arises when signal paths from different directions degrade the ability to determine the travel time of the direct path. Most problematic indoors is short multipath with delays of less than 0.5 code chips (i.e. pulses) compared to the direct signal.

The standard model for decomposition of a received multipath signal $r(t)$ is

$$r(t) = \sum_i A_i s(t - \tau_i) + z(t), \qquad (3.3)$$

where A_i is the complex amplitude of the ith path, τ_i is the delay of the ith path, $s(t)$ is the transmitted signal and $z(t)$ random noise. Based on observed data, the multipath time delays and the complex amplitudes (including phase shifts) can be estimated. The number of parameters can be reduced by solving for the complex amplitudes analytically as a function of the time delays.

One way to distinguish the direct path from an NLoS path is to move the receiver or emitter stations. NLoS paths change erratically while in motion allowing for separation and averaging, while the direct path is directly related to the motion of the object. Thus, averaging over time with a motion-tracking model is an effective way to mitigate multipath.

Another way to mitigate multipath is the use of a diversity scheme by switching to different frequency channels. Alternatively, radio signals with a large absolute frequency bandwidth such as Ultra-Wideband (Chapter 10) have been shown to be advantageous for mitigation of multipath fading (Molisch 2009).

3.3 The Basic Measuring Principles

The following basic measuring principles are common techniques for distance and angular observations which form the basis for the positioning methods described in section 3.4.

3.3.1 Time of Arrival (ToA) / Time of Flight (ToF)

The principle of ToA is based on measuring the absolute travel time of a signal from a transmitter to a receiver. The Euclidean distance between two devices can be derived by the multiplication of the signal travel time by the wave speed (i.e. speed of light in vacuum). Since the wave velocity depends on properties of the propagation medium, knowledge of the penetrated material is required. For all building materials the propagation speed depends on the square root of the dielectric constant k. For example, for glass and dry concrete $k \approx 5$, slowing down electromagnetic waves by a factor of more than 2. For ferroconcrete $k \approx 9$, resulting in a factor in travel speed velocity of one third compared to the speed of light.

ToA relies on precise synchronization of transmitter and receiver clocks, as even one nanosecond error in synchronization translates into a distance error of 30 cm if radio frequency signals are used. During Line-of-Sight (LoS), a rule of thumb is that timing can be achieved down to a fraction of the chip duration. A chip is typically a rectangular pulse of +1 or -1 amplitude, multiplied by a data sequence. ToA is particularly difficult to apply in indoor environments where multi-path conditions are common, because the autocorrelation peak in the signal referring to the LoS beam may not be resolved. The usage of a wider frequency band is a way to address this problem.

3.3.2 Time Difference of Arrival (TDoA)

Taking time differences of ToA measurements has the advantage that a possible receiver's clock bias is not relevant. Any constant time offset of a non-synchronous receiver's clock is eliminated by subtraction. In contrast to ToA, the receiver does not need to know the absolute time at which a pulse was transmitted - only the time difference of arrival from synchronized transmitters is needed. With two emitters at known locations, a receiver can be located onto a hyperboloid. A receiver's location can be determined in 3D from four emitters by intersection of three hyperboloids. With this approach, very precise synchronization of all emitters is a precondition. For GNSS positioning TDoA is a useful approach, because the drift of a low-cost receiver's clock can be eliminated while the satellites are precisely synchronized by 'GNSS time'. Conversely, a mobile emitter can be located from multiple receivers. In this configuration the infrastructure is trying to determine the location of the mobile station.

3.3.3 Round Trip Time (RTT) / Roundtrip Time-of-Flight (RToF) / Two Way Ranging (TWR)

Using RTT, also known as Two-Way Ranging (TWR), the time taken by the signal to travel from a transmitter to a receiver and back is measured. RTT avoids the need for time synchronization between the two devices, allowing its application in uncoordinated mesh networks with the advantage of low complexity and cost. As a drawback of this method, range measurements to multiple devices need to be carried out sequentially which may cause critical latencies for applications where devices move quickly.

3.3.4 Phase of Arrival (PoA) / Phase Difference (PD)

PoA uses the received carrier phase to determine the distance between two devices. In order to mitigate phase wrapping, the received signal phase is evaluated on multiple frequencies. The distance is then determined by the rate of phase change.

3.3.5 Near-Field Electromagnetic Ranging (NFER)

NFER refers to any radio technology employing near-field properties of radio waves. The principle is that the phase of an electro-magnetic field varies with the distance around an antenna. NFER has potential for range measurements in the accuracy range of 30 cm to 1 m and operating distances up to 300 m. More details can be found in Chapter 15 Magnetic Localization.

3.3.6 Angle of Arrival (AoA) / Angulation / Triangulation / Direction based Positioning

AoA information, i.e. direction of incidence, can be obtained by the use of directionally sensitive antennas. In real application, AoA is usually based on crude sector information. Kemppi et al. (2010) use a multi-antenna array for a pedestrian navigation system. A survey of methods for antenna orientation can be found in Grimm (2012). But also all camera based systems make use of the principle of AoA. Each pixel from a CCD/CMOS chip represents a distinctive horizontal- and vertical angle of incidence. If more than one AoA measurement is performed, the position can be determined by intersection of lines (i.e. resection and intersection).

3.3.7 Doppler Ranging

The Doppler ranging technique is used to observe the relative velocity between a transmitter and a receiver. If a fixed signal source is used the absolute velocity along the line of sight can be derived from the measured Doppler frequency shift. Given a known initial position and multiple Doppler frequency observations, the displacements of a mobile device can be determined.

3.4 Positioning Methods

This section gives an overview of basic principles for 2D/3D position determination from various basic measurements such as proximity, distance and angular observations.

3.4.1 Cell of Origin (CoO) / Proximity Detection / Connectivity Based Positioning

The CoO method is used to determine the position of a mobile device merely by its presence in a particular area or based on a physical phenomenon with limited range. The procedure consists in simple forwarding of the position of the anchor point where the strongest signal is received. The accuracy of CoO relates to the density of anchor point deployment and signal range. CoO is a simple positioning method used for applications with low requirements for accuracy. Examples are sensors detecting physical contact, automatic ID systems and mobile wireless positioning systems.

3.4.2 Centroid Determination

Centroid location determination involves knowledge of multiple beacon positions within the detection range and simply locates the beacon at the centroid. Alternatively the weighted centroid location can be determined, where the weights are set in function of RSSI values (see 3.2.6), distances or uncertainties of each beacon.

3.4.3 Lateration / Trilateration / Multilateration

All three terms refer to position determination from distance measurements. Usually a 2D/3D position is computed with redundancy from more than two/three distance measurements to nearby nodes. Lateration based positioning can be applied on a set of distances no matter what distance estimation method has been used. Well-known methods for distance estimation are ToA, TDoA, E-OTD, RTT, PoA and RSSI/SNR. In some literature the term multilateration is particularly used to indicate that the distances originate from TDoA. A mathematical formulation to the lateration problem and its solution can be found in Mautz et al. (2007a).

3.4.4 Polar Point Method / Range-Bearing Positioning

The polar point method uses a distance and an angular measurement from the same station to determine the coordinates of a nearby station. This method is particularly useful, because it requires measurements from only one station (under the assumption that orientation is also known). Polar point determination is conveniently used in geodetic surveying, where the position of several targets can be determined from a single set-up of a totalstation.

3.4.5 Fingerprinting (FP) / Scene Analysis / Pattern Matching

The standard quantity for Fingerprinting (FP) is radiofrequency RSSI, but FP can also be performed acoustically from audio or visually from images. Fingerprinting typically consists of two phases. First, in an off-line calibration phase, maps for fingerprinting are set up either empirically in measurement campaigns or computed analytically. In the first case, signal strengths received from fixed stations are measured at a number of points inside a building and added to a database. In the operation phase the current measured signal strength values (RSSI-tuples) are compared for the best agreement with a database. The second case of analytical model generation is used to avoid elaborate calibration measurements. Thereby, the signal strength reference values can be computed using a signal propagation model. More details on fingerprinting methods can be found in Chapter 8 WLAN / Wi-Fi.

3.4.6 Dead Reckoning (DR)

Dead reckoning is the process of estimating a position based upon previously determined positions and known or estimated speeds over the elapsed time. An inertial navigation system is the main type of sensor used. A disadvantage of dead reckoning is that the inaccuracy of the process is cumulative, so the deviation in the position fix grows with time. The reason is that new positions are calculated solely from previous positions. In literature associated with the field of indoor applications the term Pedestrian Dead Reckoning (PDR) is used as an indication that accelerometers have been attached to the body of a person.

3.4.7 Kalman Filter (KF) and Extended Kalman Filter (EKF)

The most widely used algorithm for fusion of dead reckoning positions with absolute position updates is the Kalman Filter. It is a recursive estimator of the state of a dynamic system. The classic KF seeks to maximize the conditional probability of the state **x** given the past history

$$\max_{\mathbf{x}}\{p(x_i|x_{i-1}, x_{i-2}, \dots, x_0)\}. \tag{3.4}$$

Kalman filtering is an important technique particularly used to predict positions in-between data samples for camera based tracking and pedestrian navigation based on inertial measurements. For application on nonlinear observation and state transition models the Extended Kalman Filter (EKF) is used. The required linearization of the nonlinear functions introduces the risk of divergence. Nevertheless, the EKF has become the standard algorithm for nonlinear state estimation in navigation.

3.4.8 Map Matching (MM)

Map matching algorithms – also called map measurements – combine current positioning data with spatial map data to identify the correct link on which a pedestrian (or vehicle) is travelling while improving positional accuracy. The use of maps is an economical alternative to the installation of additional hardware. MM techniques include topological analyses, pattern recognition or advanced techniques such as hierarchical fuzzy inference algorithms. A number of map matching algorithms have been assessed in Quddus et al. (2007).

3.4.9 Combination of Basic Measuring Principles and Positioning Methods

Many system architectures use multiple measuring principles which require the use of various positioning techniques. For example, in a large sensor network with a huge number of nodes, the combined use of proximity, distances and angular measurements enhances positioning performance. Recently, research focus is on the integration of heterogeneous sensor technologies that employ a combination of positioning methods and data fusion techniques.

4 Cameras

Wavelength	100 nm	1 µm	10 µm	100 µm	1 mm	10 mm	0.1 m	1 m	10 m	100 m	1 km
	UV		Infrared				Microwave			Radio	
Frequency	3 PHz	300 THz	30 THz	3 THz	300 GHz	30 GHz	3 GHz	300 MHz	30 MHz	3 MHz	300 kHz

This Chapter describes optical indoor positioning approaches where a camera is the only or the main sensor. Optical systems combined with distance or mechanical sensors are treated in Chapter 6. A literature review on optical systems has been published in Mautz and Tilch (2011) and is reprinted here for the sake of completeness.

Cameras are becoming a dominating technique for positioning which covers a wide field of applications at all levels of accuracy, with its main application area in the sub-mm domain. The success of optical methods originates from improvement and miniaturization of actuators (e.g. lasers) and particularly advancement in the technology of detectors (e.g. CCD sensors). In parallel there has been an increase in data transmission rates and computational capabilities as well as profound development of algorithms in image processing.

Optical indoor positioning systems can be categorized into ego-motion systems where a mobile sensor (i.e. camera) is to be located and static cameras which locate moving objects in images. An answer is to be found how position and rotations in a 3D world can be computed where the primary observations are 2D positions on a CCD sensor. All camera-based system architectures measure image coordinates which represent only angular information and exclusively built on the Angle of Arrival (AoA) technique. Depth information of monocular images can be obtained by making use of the motion of a camera. In this approach – known as synthetic stereo vision – the scene is observed sequentially from different locations by the same camera and image depths can be estimated in a manner similar to the stereo-vision approach. However, the baseline between sequential images needs to be determined by a complementary technique. Therefore, the system scale cannot be determined from images alone and requires a separate solution.

The transformation from image space into object space requires additional distance information. If a stereo camera system is used with a known baseline, the scale can be determined from stereoscopic images. As a drawback, the performance of a stereo

4 Cameras

camera system is directly driven by the length of the stereo baseline and therefore a miniaturization for handheld devices is not applicable.

Alternatively, distances can be directly measured with additional sensors, such as with laser-scanners (see Section 6.2.4) or range imaging cameras. The latter return a distance value for every pixel of a 320 × 240 image at a frame rate of 100 Hz. In order to determine the scale roughly, the position of the autofocus can be used.

A decisive characteristic in the system architecture is the manner how reference information is obtained. Therefore, this survey of recently developed optical navigation systems takes the primary mode of reference as criterion for categorization. An overview of the here mentioned systems and their key parameters are given in Table 4.1 at the end of this chapter.

A comprehensive survey of older works can be found in DeSouza and Kak (2002). A more recent overview of video tracking systems has been carried out by Trucco and Plakas (2006), where the main algorithmic approaches, namely window tracking, feature tracking, rigid object tracking, deformable contour tracking and visual learning are explained and 28 works of video tracking are discussed.

4.1 Reference from 3D Building Models

This class of positioning methods relies on detection of objects in images and matching these objects with a building data base (such as CityGML as shown in Figure 4.1) which contains position information of the building interior. The key advantage of these methods is that there is no requirement for installation of local infrastructure such as deployment of sensor beacons. In other words, reference nodes are substituted by a digital reference point list. Accordingly, these systems have the potential for large scale coverage without significant increase of costs.

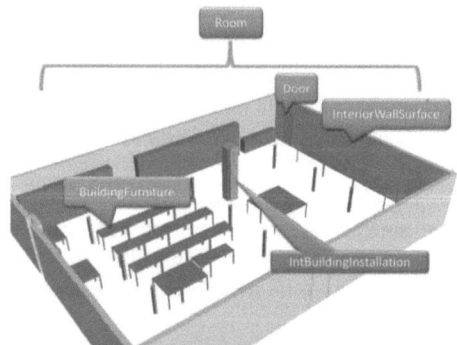

Figure 4.1 Model of a room in CityGML (Kohoutek et al. 2010)

Kohoutek et al. (2010) use the digital spatio-semantic interior building model CityGML at the highest level of detail (LoD 4) with the intention to determine location and orientation of a range imaging camera. In a first step the correct room of the camera is identified in the CityGML data base. From the 3D point cloud obtained by the range image sensor, fixed objects such as windows and doors are detected and their geometric properties can be compared with the database. The second and last step consists of dm-level fine positioning of the camera based on a technique which combines trilateration and spatial resection.

Hile and Borriello (2008) compare a floor plan with the current image of a camera phone. In a first step rough location is determined by WLAN connectivity to limit the search area. In a second step extracted features from the images are used to find feature correspondences and to compute the pose of the phone at decimeter level. Location based information can be displayed instantly.

Kitanov et al. (2007) compare image lines which have been detected in images of a robot mounted camera with a 3D vector model. The camera orientation is repeatedly computed from image sequences while the camera is in motion. An off-line optimal matching of rendered image lines and lines extracted from the camera images appears to achieve dm-level positioning accuracy. An odometer is used to stabilize system robustness.

The computer vision algorithm described by Schlaile et al. (2009) also relies on feature detection in an image sequence. Here, the computer vision module is used for complementary assistance of an integrated navigation system mounted on a micro aerial vehicle.

4.2 Reference from Images

The so-called view-based approach relies on sequences of images taken beforehand by a camera along certain routes in the building, see Figure 4.2. Thereby, the current view of a mobile camera (as shown in Figure 4.3) is compared with these previously captured view sequences. The main challenge of this approach is to achieve real-time capability. For the identification of image correspondences the computational load is particularly high since operability is assumed without deployed passive or active optical targets. Nevertheless, all systems require an independent reference source from time to time in order to control accumulated deviations.

Figure 4.2 Example of a view sequence

Figure 4.3 Current view to be compared with the view sequence

In order to navigate a humanoid robot through office buildings, Ido et al. (2009) carry out a template matching of images. In an initial phase, recording runs are captured by a camera mounted on the head of the robot. When the view sequences have been analyzed, compressed and stored as image templates, an autonomous navigation phase can commence. In this phase correlation coefficients between the image of the current view and the stored templates are computed to determine the robot's pose. First trials indicate an accuracy of 30 cm.

Sjö et al. (2009) navigate their robot based on a low-resolution camera with zoom capabilities. To approximately estimate distances to objects they use the zoom position in a first step and then carry out SLAM (Simultaneous Localization And Mapping) by computing RFC Histograms (Receptive Field Cooccurrence Histograms) from the current image and comparing them with histograms of previously captured images. In order to stabilize their SLAM method geometrically, the robot is also provided with laser scanning data.

Muffert et al. (2010) determine the trajectory of an omnidirectional video camera based on relative camera orientation of consecutive images. If there is no additional control via other references positions or directions, the recorded path drifts away from the true trajectory, similarly to inertial sensors relying merely on dead reckoning. For an acquisition time of 40 s, standard deviations of 0.1 gon are reported for yaw angles.

Kim and Jun (2008) match stored image sequences with the current view of a cap mounted wireless camera. In addition to the vision based positioning technique, deployed markers and topographical information of the indoor environment are also used to support the recognition of the location. The system is designed for augmented reality applications by annotating the user's view with additional information.

Based on the principle of optical odometry, Maye et al. (2006) develop a low-cost optical navigation device using an optical mouse sensor. The only modification to a computer mouse is a different lens tailored to higher speeds (2 m/s) and ground clearance (5 cm). In order to correct the accumulated path deviation, fixed landmarks are deployed to enable position updates. In addition, a magneto-inductive compass is employed. The reported drift for velocities of less than 2 m/s is 1% of the travelled path length.

4.3 Reference from Deployed Coded Targets

Optical positioning systems that rely entirely on natural features in images lack of robustness, in particular under conditions with varying illumination. In order to increase robustness and improve accuracy of reference points, dedicated coded markers are used for systems with demanding requirements for positioning. The markers serve three purposes for algorithmic development: a) simplification of automatic detection of corresponding points, b) introduction of the system scale, c) distinction and identification of targets by using a unique code for each marker.

Common types of targets include concentric rings, barcodes or patterns consisting of colored dots, see Figure 4.4. There are retro-reflective and non-reflective versions.

Figure 4.4 Three examples of coded targets used for point identification and camera calibration

Sky-Trax Inc. (2011) developed an optical navigation system for forklift trucks in warehouses. Coded reference markers are deployed on ceilings along the routes. On the roof of each forklift an optical sensor takes images that are forwarded to a server where they are processed centrally. The position accuracy is reported as 'between one inch to one foot'.

Mulloni et al. (2009) developed a low-cost indoor positioning system for off-the-shelf camera phones by using bar-coded fiduciary markers. These markers are placed on walls, posters or certain objects. If an image of these markers is captured, the pose of the device can be determined with an accuracy of 'a few centimeters'. Additional location based information (e.g. about the next conference room or the next session) can also be displayed.

AICON 3D Systems (2010) developed a system named 'ProCam' for industrial applications in the sub-mm accuracy range. The mobile video camera is pointed to a pre-calibrated reference point field. In order to increase robustness of the point detection, the camera emits active infrared light which illuminates the reference points. Tactile measurements are carried out manually with an integrated probe tip.

The StarGazer system of Hagisonic (2008) is tailored for robot positioning and relies on retro reflective targets mounted on the ceiling. An infrared sensitive camera observes different point patterns which are actively illuminated by an infrared light source. The point patterns are uniquely arranged on a 3 × 3 or 4 × 4 grid to identify each room, but also to determine the pose of the roving camera within sub-dm accuracy.

Lee and Song (2007) also use the principle of retro reflective targets to locate and orientate a mobile robot. Here, the corners of the triangle shaped targets are used for approximate orientation estimation and six inner sectors for unique identification. In order to achieve algorithmic robustness the difference between one image with and one without active illumination are processed. According to the stated results, the 2D accuracy is at sub-decimeter level.

Frank (2008) describes a system with the name 'stereoScan-3D' developed by Breuckmann GmbH for high precision mapping of industrial surfaces. Two mobile high resolution cameras with a fixed baseline are used to capture points with an accuracy of 50 µm. Precise positioning of the cameras is carried out in a calibration phase which consists in image capture with markers attached to the target object.

4.4 Reference from Projected Targets

Projection of reference points or patterns spares physical deployment of targets in the environment, making this method economical. For some applications mounting of reference markers is undesirable or not feasible. Optionally, infrared light can be projected to attain unobtrusiveeness to the user. In contrast to systems relying only on natural image features, the detection of projected patterns is facilitated due to distinct color, shape and brightness of the projected features. The principle of an inverse camera (or active triangulation) can be used where the central light projection replaces the optical path of a camera. The main disadvantage of active light based systems is that camera and light source require direct view on the same surface.

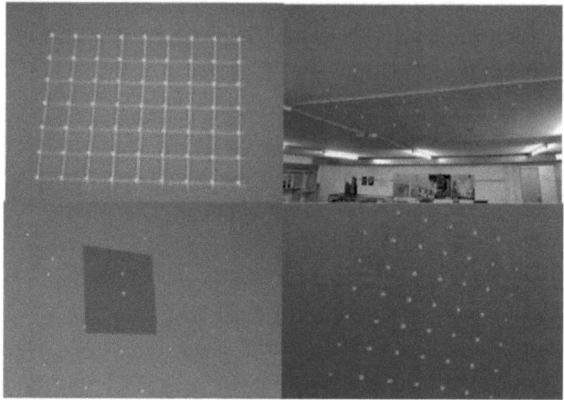

Figure 4.5 Projected reference patterns. Upper left: TrackSense Grid, upper right: CLIPS laserspots, lower left: laserspots of Habbecke, lower right: diffraction grid of Popescu

Köhler et al. (2007) built an experimental model called TrackSense consisting of a projector and a simple webcam. A grid pattern is projected onto plain walls in the camera's field of view as shown in Figure 4.5 (upper left). Using an edge detection algorithm, the lines and intersection points are determined. By the principle of triangulation – analogous to stereo vision – distance and orientation to each point relative to the camera is computed. With a sufficient number of points TrackSense can determine the camera's orientation relative to fixed large planes, such as walls and ceilings. The evaluation of TrackSense indicates that such a system can deliver up to 4 cm accuracy with 3 cm precision.

Tilch and Mautz (2010) developed CLIPS (Camera and Laser based Indoor Positioning System) with the purpose to determine the pose of a mobile camera with respect to a laser rig. Since the rig emits laser-beams from a virtual central point, it can be regarded as an inverse camera. From bright laser spots projected to any surface without any specific structure of the scene as shown in Figure 4.5 (upper right), the relative orientation between camera and laser rig can be computed. Point tracking is achieved at frame rates of 15 Hz and the accuracy of the camera position is sub-mm.

4.5 Systems without Reference

The video camera system of Habbecke and Kobbelt (2008) is based on a mobile rig of laser pointers and a fixed camera. The laser rays mounted on the rig have an arbitrary alignment without the need for a central point of intersection. In order to accomplish correct identification of the laser spots (shown in Figure 4.5 lower left), a greedy pairing algorithm is used. Via least squares minimization, the relative orientation between the camera and the rig is determined. Apart from pose determination, the system can be used for tracking and scene reconstruction. Reported accuracies indicate position deviations in the order of a few millimeters.

The laser rig of Popescu et al. (2004, 2006) is rigidly mounted to a hand-held video camera with the advantage that the relative orientation between laser source and camera remains constant. From a single laser source and a diffraction grating which acts as a beam splitter a grid of 7 × 7 laser spots is generated, see Figure 4.5 lower right. These 49 spots are located in each frame, and their 3D positions are computed by triangulation between the optical rays and the laser beams. When the camera with the laser rig is freely moved through a scene, the 3D positions of the laser spots can be used for scene modeling. The positional accuracy was reported to be better than 1 cm.

Evolution Robotics (2010) developed the indoor localization system NorthStar for navigation of shopping carts or robotic vacuum cleaners. Position and heading of the mobile unit is determined from infrared light spots, emitted from one or more infrared LED. Each mobile unit can be equipped with an infrared detector and projector to allow determination of relative orientation between mobile devices. The reported positioning accuracy is in the magnitude of cm to dm.

4.5 Systems without Reference

The purpose of systems in this class is to observe position changes of objects directly and therefore do not require external reference. The common approach is to track mobile objects with high frame rates in real-time by a single or multiple static cameras.

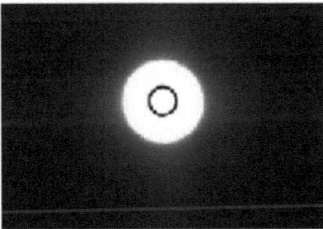

Figure 4.6 Tracked LED by DEADALUS with a circle indicating the centroid

The DEADALUS system described by Bürki et al. (2010) consists of a CCD camera clipped on a surveying totalstation. Due to the magnification of the telescope, high-precision horizontal and vertical angle measurements are possible for automated 2D monitoring of objects with high data rates. Generally, any object can be tracked, but typically illuminated targets are observed to enhance algorithmic robustness, see Figure 4.6. DEADALUS is a high-end system where the reported relative accuracy between points can reach 0.3 arc-seconds or 0.04 mm.

Boochs et al. (2010) use multiple fixed calibrated and orientated cameras to track an illuminated target mounted on the head of an industrial robot. The target body consists of a sphere with 54 self-luminous infrared LEDs to allow robust tracking from all directions. First test results of this photogrammetric tracking approach have shown 3D coordinate quality of about 0.05 mm.

Tappero (2009) suggests a low-cost system for tracking of people in an indoor environment. In order to accomplish real-time tracking using extremely cheap components the computational efficiency is optimized by detection of changes in difference images of subsequent frames. A static camera is mounted at the ceiling is able to locate people and objects within an accuracy of some decimeters.

4.6 Reference from Other Sensors

Soloviev and Venable (2010) combine vision data with GNSS carrier phase measurements for GNSS challenged environments. If less than the required number of satellites is visible, single carrier phase measurements at the accuracy level of sub-centimeters are used to support feature extraction from video images. Range measurements observed by GNSS are used to determine the system scale and facilitate feature extraction by providing image depth initialization. There is no need to determine the integer carrier ambiguities because the unknown ambiguities are eliminated by differencing between carrier phase measurements of successive positions with only a small shift in space.

4.7 Summary on Camera Based Indoor Positioning Systems

As can be inferred from the overview Table 4.1, cameras achieve accuracy levels between tens of μm and dm. Some high precision systems offer solutions for applications in surveying and industrial metrology. The covered areas of the systems presented in this chapter (excluding microscale technologies) differ between 4 m² and large room sizes or can be scaled arbitrarily. High update rates of typically more than 10 Hz allow for kinematic applications such as precision-navigation, real-time mapping and pose estimation. With the abundance of computing power and CCD sensor chips, low-cost positioning solutions are in view which have the potential to serve the mass market.

4.7 Summary on Camera Based Indoor Positioning Systems

Table 4.1 Camera systems and reported performance parameters

Name	Coordinate Reference	Reported Accuracy	Coverage	CCD Size Pixel	Frame Rate	Object / Camera Positioning	Camera Costs	Market-Maturity
Kohoutek	CityGML	dm	scalable	176 × 144	54 Hz	cam., SR 4000	9000 $	suggestion
Hile	floor plan	30 cm	scalable	640 × 480	0.1 Hz	obj., cell phone	100 $	developm.
Kitanov	vector model	dm	scalable	752 × 585	10 Hz	cam., EVI-D31	£ 245	developm.
Schlaile	edges segments	1 dm/min	scalable	752 × 582	50 Hz	cam., VC-PCC 48P	175 €	developm.
Ido	images	30 cm	scalable	320 × 40 × 4	30 Hz	cam., IEEE 1394		developm
Sjö	images / scans	sub m	scalable	320 × 240	30 Hz	cam., VC-C4R	700 €	developm.
Muffert	images	0.15 gon/min	room	1616 × 1232 × 6	15 Hz	cam., Ladybug3	>10000$	developm.
Kim	images	89 % success	scalable	320 × 240	Hz	cam, WLAN cam.		low developm.
May	images / landm.	1 %	scalable	16 × 16	2300 Hz	cam., ADNS-2051	1.35 €	developm.
Sky-Trax	coded markers	2 cm – 30 cm	scalable	-		camera		product
Mulloni	coded markers	m – dm	scalable	176 × 44	15 Hz	cam., cell phone		low product
AICON Procam	coded markers	0.1 mm	vehicle	1628 × 1236	7 Hz	cam., Procam		high product
StarGazer	coded markers	cm – dm	scalable	-	20 Hz	camera	980 $	product
Lee	coded markers	dm	36 m²	1280 × 1024	30 Hz	cam.,VX-6000	40 $	developm.
naviSCAN3D	projection	50 µm	1.5–10 m	2448 × 2048 × 2	1 Hz	obj., steroSCAN		high product
TrackSense	projection	4 cm	25 m²	640 × 480	15 Hz	obj.,cam,Pro4000	200 $	developm.
CLIPS	projection	0.5 mm	36 m²	1032 × 778	30 Hz	cam., Guppy F80	1000 €	developm.
Habbecke	projection	mm	25 m²	1280 × 960		object	1000 €	developm.
Popescu	projection	cm	25 m²	720 × 480	15 Hz	camera	1500 $	developm.
NorthStar	projection	cm – dm	36 m²	-	10 Hz	cam./obj, IR	1400 $	product
DEADALUS	none	0.04 mm	m – km	1024 × 768	30 Hz	obj., Guppy F80		high developm
Boochs	none	0.05 mm	4 m³	2000 × 2000 × 4		object		high developm.
Tappero	none	dm – m	30 m²	356 × 292	3 Hz	obj., OV 6620	20 $ US	suggestion

5 Infrared

Wavelength	100 nm	1 µm	10 µm	100 µm	1 mm	10 mm	0.1 m	1 m	10 m	100 m	1 km
	UV					Microwave			Radio		
Frequency	3 PHz	300 THz	30 THz	3 THz	300 GHz	30 GHz	3 GHz	300 MHz	30 MHz	3 MHz	300 kHz

Infrared (IR) wavelengths are longer than that of visible light, but shorter than that of terahertz radiation. Therefore infrared light is invisible to the human eye under most conditions, making this technology less intrusive compared to indoor positioning based on visible light.

System architectures for positioning based on IR signals differ to such an extent that they hardly can be summarized within a single chapter. The three general methods of exploiting infrared signals are: a) use of active beacons, b) infrared imaging using natural (i.e. thermal) radiation or c) artificial light sources.

5.1 Active Beacons

The active beacon approach is based on fixed infrared receivers placed at known locations throughout an indoor space and mobile beacons whose positions are unknown. The system architecture may include only one receiver in every room for simple room-precise localization (i.e. CoO) or one receiver with additional AoA capabilities for sub-room precision. In order to achieve meter-level precision or better, a configuration of an IR tracking system based on active beacons must include several receivers deployed in each room to disambiguate sectors of a room. Note that IR signals are unable to penetrate opaque materials, such as walls and ceilings.

One of the early and most widely recognized IR indoor positioning systems is the Active Badge System (Want et al. 1992) designed for locating people at room level. The building staff wears 'Active Badges' which emit short IR pulses with unique codes at a rate of 0.07 Hz. The signals are picked up by a network for fixed IR receivers deployed in the building interior. Since the concept of Active Badge makes use of the Cell of Origin (CoO) principle, the positional accuracy is driven by the operating range of an IR sender, which is 6 m. One major disadvantage of Active Badge is that it is not suitable for real-time applications as the positional update rate is 15 s. The largest implementation includes 200 active badges and 300 sensors.

A special application of infrared rays has been proposed by Atsuumi and Sano (2010). The system is designed to mitigate the sensor drift of integrating positioning systems by providing angular information. The azimuth angle of incidence is estimated with respect to a single beacon by the use of polarized infrared light. The beacon is composed of an infrared light source and an optical polarizing filter, which only passes light through that oscillates along a single plane. The receiver consists of a photo detector and a rotating polarizer that causes attenuation of the signal intensity depending on the horizontal angle. The phase of the time-varying signal is then translated into the angle of the polarizing plane. This allows estimation of the absolute azimuth angle with an accuracy of 2% (or a few degrees) as first experiments have shown. Note that this technique uses a special measuring principle that is not related to the AoA (Angle of Arrival) method.

5.2 Imaging of Natural Infrared Radiation

In the literature, positioning systems using natural infrared radiation are known as passive infrared localization systems. Sensors operating in the long wavelength infrared spectrum (8 µm to 15 µm, also known as the thermography region) are able to obtain a completely passive image of the surrounding world from natural thermal emissions. Thus, it is not necessary to employ active infrared illuminators or any other dedicated thermal source. Thermal infrared radiation can be used to remotely determine the temperature of people or objects without any need for wearing tags or emitters. Existing thermal detectors are thermal cameras, broadband detectors (Golay Cells), pyroelectric infrared sensors used for motion detection or thermocouples used to convert heat gradients into electricity or to measure the temperature contact-free. As a drawback, passive infrared approaches are compromised by strong radiation from the sun.

Hauschildt and Kirchhof (2010) propose a localization system based on passive thermal IR sensors to detect thermal radiation of the human skin. Their thermal IR approach uses thermophiles, which are a series of thermocouples (i.e. temperature sensor elements) with a lower resolution compared to IR cameras. Multiple sensors are placed in the corners of a room from where the angles relative to the radiation source are measured. The position is then roughly estimated via the principle of AoA. Via triangulation from multiple thermophile arrays the position of humans can be determined. First experiments indicate that positional accuracy at decimeter level is feasible, but effects of dynamic background radiation need to be studied first before the method is applicable in real environments.

Ambiplex (2011) offers the system 'IR.Loc' that includes different localization solutions based on naturally emitted heat radiation. A sensor measures the angle of incidence to a heat source. From multiple sensors mounted on the walls, the location of a heat source can be determined at a measurement rate of 50 Hz. The reported location accuracy is 20 cm to 30 cm at an operating range of 10 m. Intended applications for IR.Loc are in the area of person detection such as control of automatic doors, detection of persons in security zones or surveillance to reduce the risk of fire.

5.3 Imaging of Artificial Infrared Light

Optical IR indoor positioning systems based on active IR light sources and IR sensitive CCD cameras are a common alternative to optical systems operating in the visible light spectrum. Implementations using IR cameras are either based on active infrared LEDs such as that of Boochs et al. (2010) or based on retro reflective targets, e.g. Lee and Song (2007), AICON (2011) and Hagisonic (2008) with their StarGazer system. More details on camera systems can be found in Chapter 4.

The motion sensing device known as Kinect used for the video game console Xbox (Microsoft Kinect 2011) uses continuously-projected infrared structured light to capture 3D scene information with an infrared camera. The 3D structure can be computed from the distortion of a pseudo random pattern of structured IR light dots. People can be tracked simultaneously up to a distance of 3.5 m at a frame rate of 30 Hz. An accuracy of 1 cm at 2 m distance has been reported. The release of a Kinect software development kit has inspired several third party developments for automatic tracking, robotic guidance and gesture control and even surgical navigation. The NorthStar system developed by Evolution Robotics (2010) is also based on projection of infrared laser spots. Only two projected spots are employed by NorthStar to determine the position and orientation of a vacuum cleaner.

5.4 Summary on Infrared Indoor Positioning Systems

The spectral region of infrared has been used in various ways for detection or tracking of objects or persons. Systems based on high resolution infrared sensors are able to detect artificial IR light sources at sub-mm accuracy, whereas systems based on active beacons or those using natural radiation are mainly used for rough positional estimation or detecting the presence of a person in a room. Some performance parameters of infrared-based indoor positioning systems are quantified in Table 5.1.

Table 5.1 Positioning systems based on infrared

Name	Year	Measuring Principle	Reported Accuracy	Coverage	Target Illumination	Update Rate	Market Maturity
Active Badges (Want)	1999	cell of origin	6 m	scalable	signal transmission	0.1 Hz	product
Atsuumi	2010	polarized light	2 %	3 m	photo detector	high	demonstrator
Hauschildt	2010	angle of arrival	dm	30 m^2	natural IR radiation	-	demonstrator
Ambiplex	2011	angle of arrival	20-30 cm	10 m	natural IR radiation	50 Hz	product
Boochs et al.	2010	IR camera	0.05 mm	4 m^3	active, LED	-	development
Lee and Song	2007	IR camera	dm	36 m^2	retro reflective	30 Hz	development
AICON ProCam	2011	IR camera	0.1 mm	vehicle	retro reflective	7 Hz	product
Hagisonic - StarGazer	2008	IR camera	cm - dm	scalable	retro reflective	20 Hz	product
Evolution Robotics	2010	IR camera	cm - dm	36 m^2	IR projection	10 Hz	product
Kinect	2011	structured light	1 cm	3.5 m	passive	30 Hz	product

6 Tactile and Combined Polar Systems

Tactile and Combined Polar Systems are primarily used for industrial or surveying applications since they fulfill demanding accuracy requirements in the orders of 10^{-5} to 10^{-6}. Despite this excellent performance, which translates into 0.1 mm to 0.01 mm positional accuracy within the specified operating range of a few meters, these systems have not traditionally been considered as 'indoor positioning systems'. This is due to their high price, which is not suited for mass market applications. Research in this field is conducted by geodetic institutes and manufacturers of instruments. Published papers focus mainly on calibration or assessment of such instruments. Table 6.1 shows an overview of the instrumentation used for high-precision 3D indoor positioning and tracking.

6.1 Tactile Systems

Tactile systems are high precision mechanical instruments which measure positions by touching an object with a calibrated pointer, a so-called probe.

6.1.1 Measuring Arms

Measuring arms are manually operated devices designed for short-range indoor applications. The objective is determination of the pose of a probe or the position of a laser triangulation scanner, representing the end of an open-chain structure with q links connected to each other by $q - 1$ joints. Up to $q = 7$ links are used. High precision angle encoders in the joints measure the orientation of the next link. The coordinates of the tip are determined from the joint angles and well-calibrated lengths of the joints which form a polygonal chain. For the purpose of a detailed digital 3D acquisition and representation of an object's surface, the tip of the measurement arm can be combined with a laser scanner. The main advantages of measuring arms are portability, high precision and a high (m-1) degree of freedom which facilitates its manual operation. The operating range is typically 3 m, but can be scaled by combining measurements from multiple instrument stations. Manufacturers of measuring arms are CogniTens, Hexagon Metrology, Romer, Faro, Nikon Metrology, CAM2 and others.

6.1.2 Coordinate Measuring Machines

A Coordinate Measuring Machine (CMM), sometimes referred to as measuring robot, is stationary device for very precise 3D positioning determination. A CMM consists of a table on which the object to be measured is placed and a movable bridge with a measuring probe. The probe samples the test object along the x-, y-, and z-axes of the Cartesian machine coordinate system with extremely high accuracy in the order of 10^{-6} to 10^{-8}. Traditionally, mechanical touch probes have been used, whereas new probing systems are non-contact optical probes. Optical gratings running along the length of each axis can be used to determine the position along a scale system. The work space is limited to the size of the CMM, typically covering volumes between $(0.3\ m)^3$ and $10\ m \times 4\ m \times 3\ m$. As with measuring arms, a scanning device can be mounted on the tip for a detailed acquisition of the physical geometrical characteristics of an object. Operation of CMMs can be carried out manually or automatically by direct computer control. Manufacturers include Metris, Merlin, Leitz, Brown & Sharpe, LT, Mitutoyo, Leader Metrology and Walter.

6.2 Combined Polar Systems

Polar systems measure angles by mechanical or optical encoders or by the Time of Arrival (ToA) of a rotating beam, e.g. iGPS 6.2.6. In addition to angles, the distances to targets are also measured such that combined polar systems determine object positions from the complete 3D vector between instrument and target object.

6.2.1 Laser Trackers

Modern laser trackers are devices equipped with a telescope and a multitude of sensors, where an interferometer and two angle encoders are the key components. The laser interferometer is used for distance measurements to a tracking reflector. Nowadays absolute distance can be observed by an absolute interferometer. The spatial orientation of the emitted laser beam is determined by two angle encoders which measure horizontal and vertical positions of the rotation axes of a leveled instrument. From distance and angle observations, spherical 3D coordinates can be determined with accuracy in the order of $1 \cdot 10^{-6}$ for high-dynamic and high precision applications such as for automotive and aerospace industries. A typical operating range is 15 m, expandable to 80 m. Distances can be measured with an accuracy of $10\ \mu m + 5\ ppm\ (\mu m/m)$ and angles up to 0.001". The tracking retro reflector can be equipped with tactile devices for probing or lasers for object scanning. Modern laser trackers can determine the full pose (6 DoF) of the probe by integration of a high-speed camera system. For dynamic tracking, the system is capable of closed-loop control by following the target automatically, where the current position of the survey beam is measured by a Position Sensitive Device (PSD). Laser trackers are offered by Hexagon Metrology, Faro and Automated Precision Inc.

6.2.2 Totalstations and Theodolites

Totalstations are used for various tasks in the field of indoor and outdoor surveying engineering. Usually, a reflector prism is manually or automatically sighted through the crosshairs of a telescope. Horizontal and vertical angles are then measured visually or optically by means of precise readings of digital bar-codes on rotating discs within the instrument's alidade. Reference to the plumb line is established by a two-axis tilt sensor. In combination with an Electronic Distance Meter (EDM), relative 3D coordinates of the prism can be determined. Modern totalstations feature an Automatic Target Recognition (ATR) system and a servomotor control system for automatic prism tracking and robotic guidance. Nowadays, tracking totalstations can be used to measure real-time 3D trajectories of moving targets within an accuracy of a few millimeters (Kirschner and Stempfhuber 2008).

In contrast to totalstations, theodolites do not include a distance sensor. Two theodolites can be employed as an 'industrial measurement system' to determine 3D coordinates solely from angular measurements under the assumption that the baseline vector between the two stations is known. By the use of an industrial measurement system, a coordinate accuracy of 0.1 mm can be reached within distances of 10 m.

The maximal operating range of a totalstation is typically 2 km to 10 km, depending on the EDM and the prism. Despite these performance parameters, totalstations are not associated with indoor positioning, due to high hardware costs, hardware size and need for LoS between instrument and target. Leading manufacturers of totalstations are Trimble, Hexagon Metrology (Leica Geosystems), Topcon and Sokkia.

6.2.3 3D Disto
Leica Geosystems (2011) has developed a low-cost totalstation for indoor applications. The so-called '3D Disto' combines a laser projector, a laser distance meter, horizontal and vertical goniometers, tiltmeters and a camera within one instrument which can be placed on a table with a self-leveling function. The laser projector is used to project laser patterns on the walls for setting out. The 3D Disto is designed to assist plumbers, building managers or roofers.

6.2.4 Laser Scanners
Terrestrial laser scanners are based on a non-contact ranging technology for 3D point measurement and 3D point cloud acquisition. In combination with readings at a horizontal and a vertical circle, primarily polar coordinates of the measured points are determined and then transformed into Cartesian coordinate system. This process generates a so-called point cloud of the scene. The main principles for distance determination are optical triangulation, phase and pulse measurements. From a single instrument set up close range scanning allows capturing an area less than 100 m and up to 3 km if a long range scanner is used. Scans from multiple stations can be combined via registration, making laser scanning a scalable technique.

Manufacturers are – amongst others – Sick, Faro, Zoller + Fröhlich, Riegl, Callidus, Mensi, Optech, Leica, VisImage and Mensi. An overview of terrestrial laser scanning is provided in Fröhlich and Mettenleiter (2004).

Khoshelham (2010) presents an indoor localization method for terrestrial laser scanners via matching planar objects (such as walls) in multiple scans. If at least three correspondences of intersecting planes can be found in two scans, relative scanner positions can be determined with an accuracy of a few centimeters. This method is useful for the process of fine registration of multiple scans, but it is no real-time method for indoor positioning due to long acquisition times required for capturing large 3D point clouds.

Kokeisl (2011) offers a laserscanning based navigation and collision avoidance system for industrial robots. The position reference is obtained relative to the environment. From automatic scans the environment is mapped and the navigation path is adjusted accordingly in case of obstacles. An accuracy of 1 cm at speeds of 1.5 m/s is reported.

6.2.5 Laser Radar

Laser radar, known as Frequency Modulated Coherent Laser Radar (FM CLR) uses the principle of laser scanning, by sending an infrared laser beam to a target object where the signal is reflected and a fraction of it is returned. In contrast to laser scanning, a part of the beam travels through a calibrated optical fiber for reference. The two returning beams are superimposed by non-linear mixing (heterodyne detection). This technology allows for precise measurement of the absolute distance. Up to 2000 points per second can be measured. Laser radars can be used at a surface reflectivity of less than 1 % and operate independent of lighting conditions. The main application is the inspection of large volume parts for aerospace and shipbuilding industries. A manufacturer is Nikon Metrology and a retailer is Quamt Engineering.

6.2.6 iGPS

iGPS (indoor Global Positioning System) is a laser-based 3D measurement system which can be used for high precision industrial measurements. Its name 'iGPS' is misleading, because the measuring principle is different from its space-based counterpart GPS. iGPS consists of two or more static transmitters which continuously send out two rotating fan-shaped laser beams and a reference infrared pulse (see Figure 6.1). Based on Time Difference of Arrival (TDoA) between the three signals, the relative horizontal and vertical angles with respect to a receiver are determined. In order to establish a time-dependency between the arrival of the laser planes and the vertical angle, the rotation axis of the laser planes are tilted by an elevation angle of α_{el} = 30°. The first laser plane follows the second at a 90° angle delay (one plane is tilted 30° to the left, the other 30° to the right). The horizontal angle to a receiver is estimated from TDoA between a laser plane and the reference pulse. A receiver's 3D position is determined from two angles of at least two orientated transmitters whose coordinates have been determined in a prior set-up phase. One can say that iGPS is based solely on triangulation, or more precisely on spatial resection. The system scale is introduced from the fixed length between two receivers as shown in Figure 6.2. Two receivers mounted on a fixed stick (the so-called vector-bar) allow for determination of the stick's full pose (6 DoF).

The manufacturer (Nikon Metrology 2011) states that the accuracy of 3D positions is 0.2 mm and a typical measurement footprint based on 4 to 8 transmitters is 1200 m^2. A more detailed assessment of the system is provided by Schmitt et al. (2010).

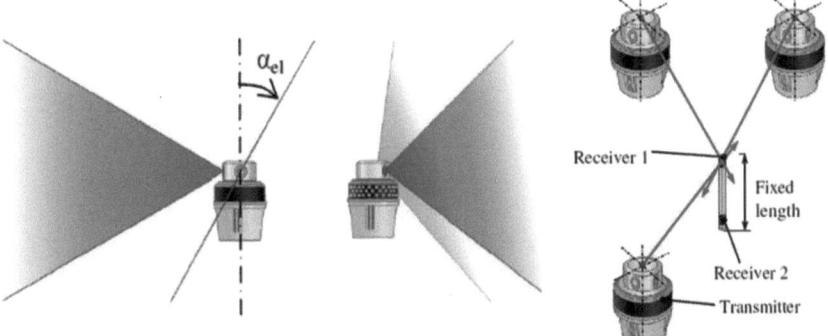

Figure 6.1 Laser planes of iGPS. Graphic by Schmitt et al. (2010), origin from Nikon.

Figure 6.2 iGPS scale determination, by Schmitt et al. (2010), origin from Nikon

Due to the rotational speed of the laser planes (40 Hz to 50 Hz) and an updated rate for receiver positions at 40 Hz, iGPS can be used for kinematic tracking. Depenthal (2010) assessed the performance under kinematic conditions and concluded that the dynamic mode enables real-time tracking of receivers at velocities of 3 m/s and 3D position deviations of less than 0.3 mm.

6.3 Summary on Tactile and Combined Polar Systems

Tactile and combined polar systems provide µm to mm level accuracy, outperforming all other technologies presented in this work. Due to high instrument costs and limitation to a single room at a time, these systems are rarely applied to navigation, but are essential tools for industry, surveying and 3D modeling. The key performance parameters of these instruments used in geodesy and metrology are quantified in Table 6.1.

Table 6.1 Tactile and combined polar systems used for geodetic and industrial applications

Device	Typical Accuracy (m)	Measuring Range (m) or Area (m²)	Measuring Principle	Application	Typical Hardware Costs	Market Maturity
measuring arm	15 µm + 10 ppm	3 m	polygonal chain	object inspection	> 30.000 €	product
CMM	2 µm + 0.4 ppm	10 m × 3 m	mechanical, interferometry	automotive, inspection	size depend.	product
laser tracker	10 µm + 5 ppm	80 m	distance & angular meas.	automotive, aerospace	> 110.000 €	product
totalstation	2 mm + 5 ppm	> 2000 m	distance & angular meas.	surveying, multipurpose	10.000 €	product
laser scanner	3 mm + 5 ppm	< 1000 m	distance & angular meas.	3D modeling	50.000 €	product
3D Disto	2 mm	50 m	distance & angular meas.	setting out, plumbing	8.000 €	product
laser radar	15 µm + 5 ppm	120 m	distance & angular meas	aerospace, industry	250.000 €	product
iGPS	0.2 mm	1200 m²	TDoA, resection	industry, metrology	> 60.000 €	product

7 Sound

In contrast to electromagnetic waves (Chapters 8-13), sound is a mechanical wave that is an oscillation of pressure transmitted through a medium. Positioning systems use the air and building material as propagation media.

An established concept using sound waves for positioning is that locations of mobile nodes on robots or tags worn by human users are determined via multilateration based on distance measurements to static nodes mounted permanently at the ceiling or walls. Most systems use ultrasound but as an alternative, audible sound can be used (see Section 7.2).

7.1 Ultrasound

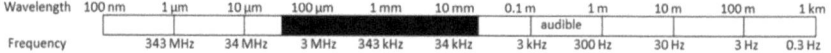

The relative distance or range between two devices can be estimated from Time of Arrival (ToA) measurements of Ultra Sound (US) pulses which travel from an US emitter to an US receiver. In contrast to radio waves, the US ToA operating range is 10 m or less due to the specific decay profile of the airborne acoustic channel. Doubling the distance causes the signal's sound pressure level to attenuate by 6 dB due to radial intensity attenuation and absorption which translates to an inverse quadratic attenuation in 3D space.

An estimation of the emitter's coordinates is possible by multilateration from three or more ranges to fixed receivers deployed at known locations. Systems operating on such architecture are known as active device systems. The alternative system architecture consists of a reverse signal flow with multiple static emitters at known locations and one or more mobile passive devices which receive the signal. In the literature ultrasound receivers are denoted as listeners and broadcasting nodes as beacons.

In order to avoid the need for time synchronization of the mobile unit with the fixed nodes and to enable ad hoc localization in unprepared environments, the Time Difference of Arrival (TDoA) method is used instead of ToA. Each beacon broadcasts an RF (Radio Frequency) signal together with an ultrasonic pulse in order to trigger nearby receiver nodes. The range r between beacon and listener can then be derived by the difference of arrival times Δt between the RF and the US signal,

$$\Delta t = \frac{r}{v_{US}} - \frac{r}{v_{RF}},\qquad(7.1)$$

where $v_{US} \simeq 344$ m/s is the speed of sound in the air and $v_{RF} \simeq 3 \cdot 10^8$ m/s the speed of light. Because RF travels about 10^6 times faster than ultrasound, the travel time of the RF signal can be neglected. The most critical influence on the sound speed v_{US} is the temperature T. The dependency is

$$v_{US} = (331.3 + 0.606 \cdot T)\, \text{m} \cdot \text{s}^{-1},\qquad(7.2)$$

where T is the absolute temperature in Celsius (°C). At room temperatures this effect causes about 0.2 % systematic range deviation per degree Celsius. For a typical maximum range of 10 m, a change of 1° C in the temperature causes a deviation in the range estimation of 2 mm. Therefore, most ultrasound systems include sensors for automatic temperature compensation. Nevertheless, the true air temperature along the path between transmitter and receiver remains unknown. Minor influences on the speed of acoustic sound are the air pressure, the CO_2 content and the sound amplitude. Temporal changes in the speed of sound can also be compensated by taking into account calibration measurements between known nodes.

Applications of positioning systems based on ultrasound are indoor tracking of people and mobile devices, but these systems are rarely used in outdoor environments due to the following three reasons: Temperature gradients are larger outdoors and therefore more complicated to model or compensate. Secondly, wind degrades the system precision significantly. Thirdly, the typical limitation for distance measurements of about 10 m is not practical for outdoor use. Although ultrasound systems are scalable by adding further fixed nodes, the deployment of nodes in outdoor environments is problematic due to the absence of ceilings or nearby walls.

Further challenges of ultrasound systems are mitigation of multipath propagation (Mautz and Ochieng 2007) and detection of heavily Doppler-shifted signals (Alloulah and Hazas 2010). A practical challenge arises from the task of minimizing the battery drain that wireless ultrasonic systems usually face. Therefore power-saving techniques have been implemented. Another negative influence on ranging accuracy is the near-far problem of transmitters (see Chapter 12) and the directivity of the transducers (Hazas and Hopper 2006).

In order to make a listener-beacon system functional, the coordinates of the static nodes need to be known in advance. Therefore the static nodes need to be surveyed by another positioning system. Because the accuracy of the node coordinates should be at least as good as the ultrasound positioning system itself, time-consuming manual positioning methods such as totalstation measurements are required – delivering a positioning accuracy of 5 mm to 10 mm (1 σ). In order to avoid the use of a second positioning system for static node localization, Mautz and Ochieng (2007) have implemented an auto-localization algorithm. As shown in Figure 7.1, a dynamic sensor node is moved slowly to various locations in a room while permanently collecting ToA ranging data to at least four beacon nodes mounted at the ceiling. The mobile and the static node positions are unknown. Even the inter-beacon ranges may not be available due to obstacles between them.

The described scenario nevertheless allows creation of a rigid distance network based on local coordinates. The deployment of static nodes in the four corners of a room allows the setup of a meaningful local coordinate system orientated along the orthogonal walls. The redundancy of such an auto-localization problem in 3D is given by

$$\text{reduncancy} = n_r - 3\,(n_s + n_m) + 6, \tag{7.3}$$

where n_r is the number of observed ranges, n_s the number of static nodes and n_m the number of mobile positions. If direct line of sight conditions allow all combinations of ranges to be obtained, then $n_r = n_s \cdot n_m$ holds.

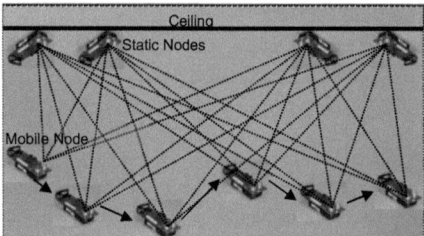

Figure 7.1 Auto-localization of static nodes using a mobile sensor

In Figure 7.1, the mobile node has collected 4 ranges to 4 static nodes at 6 different locations. For the 3D case, there are $3 \cdot (n_s + n_m) = 30$ unknown coordinates and $n_r = n_s \cdot n_m = 24$ range measurements creating a network with a redundancy of 6. But taking into account the 6 degrees of freedom (i.e. 3 translations and 3 rotations) for the rigid 3D distance network, the solution has no redundancy for this case. Mautz and Ochieng (2007) have shown that zero or low redundancy of the network causes the auto-localization algorithm to fail under real field conditions. With too low redundancy, gross errors in the position estimation are likely to remain undetected due to the existence of outlier observations, bad geometric constraints and errors caused by linearization of the objective function. To obtain the beacon coordinates the following procedure can be carried out:

a) Stepwise movement of the mobile node in a room while collecting range measurements.
b) Grouping of the ranges at m mobile positions according to their time stamps.
c) Gross error detection by finding sudden jumps in the range measurements and testing triangle conditions.

d) Creation of a distance matrix between all network positions with size $(n_s + n_m)$ by $(n_s + n_m)$, see Figure 7.2.
e) Filling the gaps of the distance-matrix by interpolation. This step establishes rough approximations for all inter-nodal ranges.
f) Setting up a local coordinate system based on the inter-nodal ranges of four nodes (preferably static nodes).
g) Computation of all coordinates based on Multi-Dimensional Scaling (MDS), a localization method which transforms a distance network into geometric embedding. Given a set of pairwise distances in presence of large range measurement noise, the location approach using MDS has outperformed proximity based algorithms (Shang et al. 2004). If the range measurements are of good quality, positions can be determined by multilateration.
h) Refinement of the coordinates by geodetic network adjustment. When applied to real data, the network adjustment may not be carried out straight away, because it requires good approximate values of the unknown positions. In order to avoid a failure of the network adjustment, a heuristic optimization method can be carried out which avoids linearization by directly using the original observation equations. The heuristic method improves approximate positions for the input of the network adjustment – typically from meter to centimeter level. An insight into heuristic methods is given in Mautz (2002).

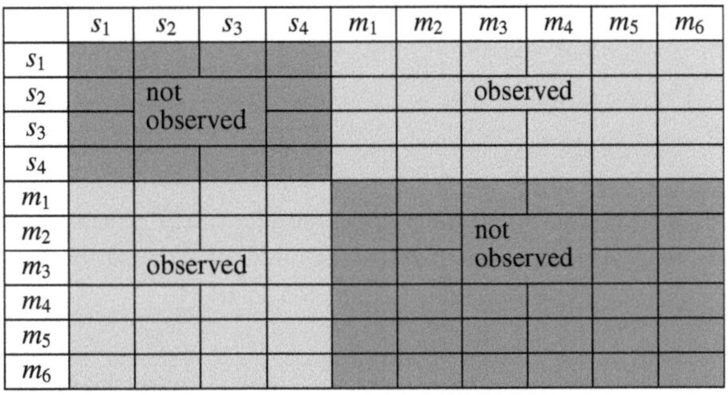

	s_1	s_2	s_3	s_4	m_1	m_2	m_3	m_4	m_5	m_6
s_1										
s_2			not observed				observed			
s_3										
s_4										
m_1										
m_2							not observed			
m_3			observed							
m_4										
m_5										
m_6										

Figure 7.2 Distance matrix of the example with 4 static nodes and 6 mobile positions

When the auto-localization procedure has been completed, the coordinates of the static nodes are available in a local system. An over-determined auto-location setup allows determining quality indicators of the coordinates.

7.1.1 Active Device Systems

In an active device system, mobile devices actively transmit signals. Therefore, the transmission of ultrasonic pulses needs to be well scheduled among the mobile devices. The disadvantage of an active device system is insufficiency in scalability, i.e. if too many users with ultrasonic devices gather in a room, the chance of signal overlap is increased.

A well-coordinated scheme for pulse transmission between the devices may solve the problem but at the cost of a reduced measurement rate.

Ward et al. (1997) describe an ultrasonic system named 'Active Bat' which can be regarded as pioneer work in the development of broadband ultrasonic positioning systems. Active Bat consists of roaming transmitters attached to the user and fixed ultrasonic receivers mounted on the ceiling. Each transmitter's position is determined by performing ToF multilateration. The Active Bat system also deduces direction information by attaching multiple transmitters on the mobile object. However, Active Bat employs a centralized system architecture and requires dense deployment of precisely positioned ultrasonic receivers. The 3D accuracy of a synchronous transmitter is 3 cm in 95 percent of cases. A large demonstration system consisting of 720 receivers and up to 75 mobile tags has been deployed in an office space of 1000 m^2.

In order to mitigate the problem of multiple user detection which active device systems usually face, Alloulah and Hazas (2010) implemented a Code Division Multiple Access (CDMA) despreader on a broadband ultrasonic signal. CDMA allows simultaneous transmission of different data streams at the same communication channel.

Sato et al. (2011) developed a range measurement technique called Extended Phase Accordance Method (EPAM) which showed a standard deviation of less than 1 mm in a laboratory experiment. However, the overall position accuracy of the real implementation with tags attached to a body is reported as 4 cm.

Sonitor IPS (2011) is a commercially available ultrasonic positioning solution which has been installed on a large scale in several hospitals for the purpose of patient and equipment tracking at room or sub-room level accuracy. Motion activated tags worn by the user transmit ultrasonic signals with unique identification codes to wall mounted receivers which process the signal and transfer relevant information to a central server via LAN or WLAN. The maximum range is specified as 18 m.

Hexamite (2011) offers ultrasonic positioning systems with various architectural set-ups. According to the company the products are supposed to provide accuracy of 9 mm and precision of 1 mm for distances up to 14 m. A standard evaluation package contains three receivers and two transmitters. Intended applications are robotic guidance, automation and manufacturing as well as 3D film studio assistance.

7.1.2 Passive Device Systems
Passive systems rely on permanently installed transmitters which broadcast ultrasound signals to the receiving devices. Such an architecture has the advantage that the privacy of the users can be kept. Mobile passive devices only receive signals and do not transmit anything. Therefore the localization is exclusively carried out onboard of the device, without any need for network interaction. The number of users can be scaled without any risk of signal overlapping.

7.1 Ultrasound

Priyantha (2005) describes in his thesis the passive 'Cricket' system developed by the MIT Laboratory for Computer Science. The Cricket system is pioneer work dating back to the year 2000. A 3D positioning accuracy of 1 cm to 2 cm can be reached indoors within a maximum range of 10 m. Real-time tracking is generally possible with an update-rate of 1 Hz. A hardware unit can be programmed either as a static beacon or mobile listener. Due to its open system architecture, Cricket has been used as a research platform all over the world and inspired many applications.

A modern, well-designed version of a passive system is presented in Schweinzer and Syafrudin (2010). Although the precision of TDoA distances is better than 0.7 mm (empiric 1-σ standard deviation of repetitive measurements), the absolute accuracy of a located listener is reported as 1 cm.

Jiménez et al. (2009) present the passive device system christened 3D-LOCUS based on TOF multilateration and delivering 1 cm accuracy. In order to apply pseudo-random digital codes they have chosen to use acoustic transducers which operate mainly in the audible region below 25 kHz. In order to serve the intended application area of archeological surveying, special pointing rods with receivers at both ends have been developed.

7.1.3 Echolocation

Similar to biosonar used by several animals such as bats, it is possible to emit sound pulses by a transmitter to the environment and make use of the returned echoes to locate and identify objects or even determine the transmitter's location. The principle of binaural (i.e. two-receiver based) localization is illustrated in Figure 7.3. Echolocation systems have the advantage that they can operate without the need for beacons or tags.

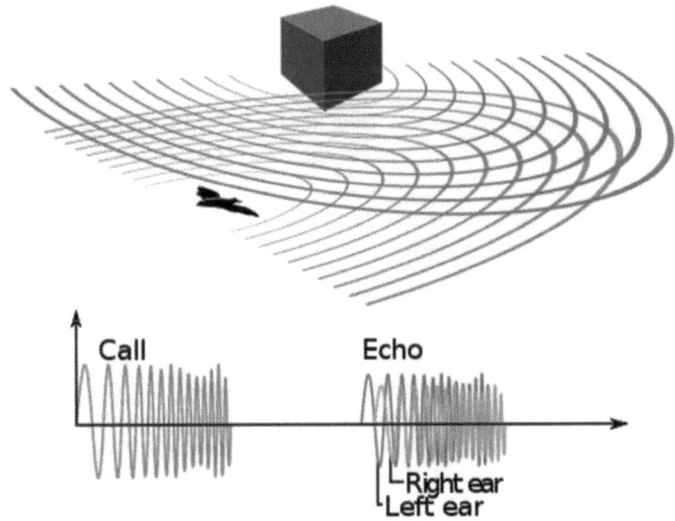

Figure 7.3 Principle of echolocation, graphic by Petteri Aimonen (2009)

Reijniers and Peremans (2007) propose a biomimetic sonar system based on distance and bearing estimation of reflectors using a central transmitter and two receivers pointing in slightly different directions. A sweeping pulse of 3 ms duration in the frequency range between 31 kHz and 59 kHz is transmitted and the echoed signal is evaluated in the time- and frequency domain. First results indicate an uncertainty of the angles of a few degrees and positions of a few decimeters.

Wan and Paul (2010) implemented an experimental setup with 6 ultrasonic wall-mounted transmitters in order to track persons without the need for body-worn tags. An US signal is reflected at a person's body causing an echo which is analyzed for estimation of the distance between the person and the transmitter. From multiple sensors it is possible to determine the person's 2D position with an accuracy of better than 0.5 m.

7.2 Audible Sound

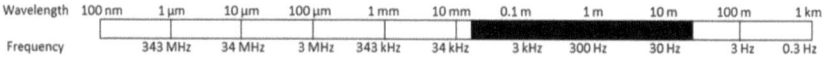

A small fraction of system approaches makes use of sound waves in the audible spectrum. The key idea is to make the system easily deployable using sound cards of standard devices. Apart from the task of avoiding annoyance to the user, other challenges arise from low bit rates and delays caused by the soundcards.

Filonenko et al. (2010) propose the use of inbuilt mobile phone speakers to generate simple sine tone ultrasonic signals in the near audible spectrum of 17 kHz to 22 kHz. Their goal is to implement ToF trilateration positioning using unmodified hardware on ordinary mobile phones.

Mandal et al. (2005) locate standard mobile devices using audible sounds at a frequency of 4 kHz. To minimize intrusiveness, positioning is carried out on request only. After calibration of these delays, the TOF multilateration approach achieves 3D accuracy of 60 cm (2 σ) with a range of 7 m.

7.3 Summary on Sound Systems

Sound systems are used for various tracking applications at cm-level accuracy. The strong decay profile of acoustic waves causes sound systems to be limited to a 10 m operating range if not scaled with additional node deployment. Time synchronization is simple due to the relatively slow speed of sound. The drawbacks are frequency changes due to the Doppler shift and a strong temperature dependency. In contrast to other technologies, the performance parameters of various sound systems are relatively similar as shown in Table 7.1. Due to existing NLoS conditions and multipath

7.3 Summary on Sound Systems

propagation in indoor environments, the development of reliable sound systems at cm-level or better remains a substantial challenge.

Table 7.1. Localization systems using sound waves and reported performance parameters.

Name	Year	Reported Accuracy	Active or Passive Devices	Transmitters	Receivers	Tags or Tag Free	Carrier Frequency	Update Rate	Principle	Application	Market Maturity
Active Bat	1997	3 cm	active	75		720 tags	40 kHz	5 Hz	multilateration	smart tracking	prototype
Alloulah	2010	3 cm	active	5		4 tags	20-50 kHz		multilateration	AAL, monitoring	demonstr.
Sato	2011	4 cm	active	5		4 tags	40 kHz	10 Hz	multilateration	human motion	demonstr.
Sonitor	2011	subroom	active	>99		>99 tags	35-40 kHz	5 Hz	disclosed	hospitals, mines	product
Hexamite	2011	0.9 cm	active	>99		>99 tags	40 kHz	30 Hz	multilateration	3D studio	product
Cricket	2005	1-2 cm	passive	20		>99 tags	40 kHz	1 Hz	multilateration	smart tracking	product
Schweinzer	2010	1 cm	passive	5-6		>99 tags	35-65 kHz	10 Hz	multilateration	WSN	demonstr.
Jiménez	2009	1 cm	passive	>5		>8 tags	<25 kHz	10 Hz	multilateration	archeology	prototype
Reijiniers	2007	decimeter	echo	1		2 free	31-59 kHz	low	echolocation	robot guidance	study
Wan	2010	0.5 m	echo	6		1 free	40 kHz	2 Hz	body reflection	person tracking	study
Filonenko	2010	< 1m	active	1		1 tags	17-22 kHz	1 Hz	multilateration	LBS, guidance	proposal
Mandal	2005	60 cm	active	1		6 tags	4 kHz	request	multilateration	LBS, malls	demonstr.

8 WLAN / Wi-Fi

Wavelength	100 nm	1 μm	10 μm	100 μm	1 mm	10 mm	0.1 m	1 m	10 m	100 m	1 km
	UV		Infrared				Microwave			Radio	
Frequency	3 PHz	300 THz	30 THz	3 THz	300 GHz	30 GHz	3 GHz	300 MHz	30 MHz	3 MHz	300 kHz

WLAN (Wireless Local Area Networks, IEEE 802.11 standard; 'Wi-Fi' is used interchangeably or as a superset of IEEE 802.11 and denotes the registered trademark of the Wi-Fi Alliance) can be used to estimate the location of a mobile device within this network. The use of WLAN signals is a tempting approach, since WLAN access points are readily available in many indoor environments and it is possible to use standard mobile hardware devices. The range of 50 m to 100 m which is typically covered by WLAN outreaches that of Bluetooth or RFID. Another advantage of using WLAN is that line of sight is not required. ToA, TDoA or AoA methods are less common in WLAN due to the complexity of time delay and angular measurements. The most popular WLAN positioning method is to make use of RSSI (Received Signal Strength Indicators) which are easy to extract in 802.11 networks and can run on off-the-shelf WLAN hardware. Therefore, WLAN positioning systems have become the most widespread approach for indoor localization, see Table 8.1 for an overview. In general, the use of RSSI in combination with WLAN can be subdivided into four strategies: Propagation modeling, Cell of Origin (CoO), Fingerprinting (FP) and multilateration.

8.1 Propagation Modeling

Modeling of signal propagation is considered as analytical fingerprinting in the literature, but treated separately here. For the purpose of analytical determination of Received Signal Strength Indicators (RSSI), different propagation models are used. A major technical difficulty using RSSI arises from the fact that RSSI values depend to a large extent on the propagation environment, making it very difficult to set up suitable propagation models which describe the relationship between RSSI values and receiver position in real indoor environments. Therefore fingerprinting methods (see **8.3**) which rely on simple comparison of empirical measurements without application of a theoretical model have become the more favorable method compared to analytical modeling. However, analytical modeling can be used in combination with empirical fingerprinting. The better a theoretical model can predict the measurements, the less calibration measurements are required in an offline stage.

A basic propagation model is the radial symmetric free-space path loss model (3.1). The purpose of this model is to derive the distance between a radiating source and a receiver by exploiting the attenuation of RSSI with distance. However, this simplistic model is rarely applicable in indoor settings where the signals do not attenuate predictably with the distance due to shadowing, reflection, refraction and absorption by the building structures. The Multi Wall Model (MWM) is a modified version of the path loss model which takes into account the wall being crossed on the direct path. It requires the thickness of walls and the dielectric properties of each wall as input. Since MWM considers only the direct path which does not necessarily correspond to the strongest path, ray tracing algorithms have been proposed which take into account all possible paths and sum up each individual contribution. In order to avoid the computational burden intrinsic to ray tracing Parodi et al. (2006) used the Dominant Path Model (DPM) which takes only into account the strongest path which is not necessarily identical to the direct path. An accepted propagation is the Indoor Path Loss Model (3.2.8) which has been validated by Chrysikos et al. (2009). However, the success of modeling WLAN propagation indoors is limited due to the interplay of highly unpredictable multipath fading.

8.2 Cell of Origin

This is a straightforward method suitable for applications with requirements for positioning accuracies of 50 m or more. The WLAN access point generating the highest RSSI value at the mobile device is identified and the user position is assumed to have the same coordinate position as the access point.

8.3 Empirical Fingerprinting

The empirical fingerprinting approach requires – as a major drawback – a previous set of calibration measurements in an offline stage (also denoted as the calibration phase) where the RSSI are observed at various locations in the building and stored together with ground-truth locations in a database known as a 'radio map'. Each entry of the database is denoted as fingerprint **f** and consists of the ground-truth position (x, y, z) and the RSSI vector **c** containing the calibration measurements to access points. The database is then used in the operational online stage to estimate the position of a mobile device by correlating all current RSSI measurements $\mathbf{r}_t$ received at time t from n different access points with the RSSI values **c** of fingerprint location **f** stored in the database. A common method is to determine the Euclidean distance

$$d(\mathbf{r}_t | \mathbf{f}(x,y,z,\mathbf{c})) = \sqrt{(\mathbf{r}_t - \mathbf{c})^2} \qquad (8.1)$$

in signal space, where the vector of RSSI differences $(\mathbf{r}_t - \mathbf{c})$ is of dimension n, and n is the number of access points. Then, the Euclidean distances d_j for all m fingerprint locations are determined and stored in the distance vector **d**. In the most straightforward approach, the minimum Euclidean distance d_{min} = min (**d**) to a calibration point is taken as the current location. Alternatively, the coordinates of the observation point can be computed by a weighted mean of the calibration points, where the weights are set reciprocal to their distances d_j. For example, the weighted mean for the coordinate x reads

$$x = \left[\sum_{j=1}^{m} \frac{1}{d_j}\right]^{-1} \sum_{j=1}^{m} \frac{x_j}{d_j}. \qquad (8.2)$$

In the so called probabilistic approaches, the correlation values

$$cor_j = cor\left(\mathbf{r}_t | \mathbf{f}_j(x,y,z,\mathbf{c})\right) = \sqrt{\frac{\sum_{i=1}^{n} \mathbf{d}_i^2}{n}}, \qquad (8.3)$$

for all j fingerprint locations are compared. Based on the correlation value cor_j the probability that the current RSSI observations $\mathbf{r}_t$ have been received at the same position as the fingerprint $\mathbf{f}_j$ can be computed. Since records at an observation point contain more than one epoch, fuzzy logic methods have been applied to make use of the additional information (Teuber and Eissfeller 2006).

Fingerprinting performance can reach meter-accuracy, depending on the number of base stations per m² and the density of calibration points where the fingerprints are taken. Even if there are no changes to the environment, recorded RSSI fluctuates in time. In order to eliminate the high-frequency deviation of attenuation in the signal (known as fast fading term) the RSS values are to be averaged over a certain time interval up to several minutes at each fingerprint location. Owing to the substantial cost incurred by building the radio map, the fingerprinting approach may be prohibitive for many applications. In order to make the creation of radio maps economically efficient, Bolliger (2008), Park et al. (2010) and Hansen et al. (2010) draw on active user participation relying on the contribution of end users by marking their location on a floor plan while recording the fingerprints. The amount of prior fingerprint measurements can also be minimized by using mutual measurements between the access points to set up an initial propagation model which is refined by incorporating additional parameters via online learning (Parodi et al. 2006).

The main drawback of WLAN fingerprinting systems is that changes in the environment such as the moving of furniture in offices, open/closed doors or even people may necessitate recalculation of the predefined signal strength map. Chen et al. (2005) quantified these dynamic changes empirically and employed RFID sensors together with an online calibration scheme to build multiple radio maps under various environmental conditions. Hansen et al. (2010) conclude from long-term measurements over a period of two months that static radio maps cannot be used for room identification even in modestly dynamic environments and therefore recommend dynamically adapting algorithms. Also the orientation of the devices contributes to a change in the RSSI values (Xiang et al. 2004) and the influence from the user's body (Gansemer et al. 2010). King et al. (2006) observed a decrease in the reception power of 15 dBm from the blocking

8.3 Empirical Fingerprinting

effect of a human body. Even the humidity level causes changes in the RSSI, due to the WLAN operation band of 2.4 GHz being a resonant frequency of water (Chen et al. 2005).

WLAN hardware and network protocols have not been designed for positioning. As a consequence, another weakness of WLAN fingerprinting arises when the method is applied to users with different WLAN chipsets. RSSI values collected from different chipset vendors differ in the RSSI definition and are therefore difficult to compare (Koski et al. 2010). Vaupel et al. (2010) determined offsets in the RSSI values of 16 dBm between different handheld WLAN devices. Their calibration approach for RSSI offsets takes into account different distances, orientations and multiple access points. After calibration, the same localization accuracy could be achieved independent of the hardware used.

The indoor localization system RADAR is a pioneer work in WLAN fingerprinting. Bahl and Padmanabhan (2000), who developed the system, used the K-Nearest Neighbor (KNN) method in signal space for location of a person and achieved a median deviation of 5 m using 3 access points covering about 1000 m^2. Apart from the node density, their findings include that the performance depends on the number of data points taken and the orientation and speed of the user.

With extensive pre-calibration (1 m grid of calibration points, 110 RSSI measurements at each calibration point in 8 different orientations) King et al. (2006) report an average deviation of 1.6 m. Their approach includes the use of an additional digital compass to detect the orientations of the users to handle blocking effects from the user. The static radio map does not model environmental changes.

The fingerprinting method of Gansemer et al. (2010) developed an adapted Euclidean distance model which takes into account a dynamically changing environment with changing sets of received base stations. Therefore, the distances in the model are normalized according to the number of stations received.

In order to make their WLAN fingerprinting system economically feasible for campus-wide deployment, Gallagher et al. (2010) simplify the calibration phase, while accepting room-level accuracy. Their system also integrates GNSS measurements and aims to provide indoor and outdoor localization for students and staff based on smart phones.

8.3.1 Commercial WLAN Fingerprinting Systems

The XPS WLAN fingerprinting system offered by Skyhook© is a software-only, server based solution which allows determining a mobile position in dense urban areas. Skyhook (2011) builds up and maintains a global database of WLAN access points by collecting RSSI data from specially equipped cars. GNSS and cell tower ID's (CoO) are also used. Since the fingerprinting is done by Skyhook, positioning can commence immediately after software installation. The system determines a device location with 10 to 20 meter accuracy in outdoor environments. The positioning performance for indoor environments was 30 m to 70 m average accuracy in a test measurement at

8 WLAN / Wi-Fi

availability of 50 % of the time (Gallagher et al. 2009). The system is designed to provide efficient positioning information on a large scale.

Also commercially available is the Ekahau© Real-Time Location System, based on the combination of RSSI WLAN fingerprinting and track history. In contrast to Skyhook, dedicated WLAN tags are to be deployed and the RSSI database is to be generated by the system providers themselves. According to the Ekahau© specifications, WLAN tags can be tracked at 1 to 3 meter accuracy. Gallagher et al. (2009) state an average accuracy of 7 m for indoor environments.

8.4 WLAN Distance Based Methods (Pathloss-Based Positioning)

Information regarding the distance from a transmitter is contained in the arrival times and the amplitudes of the received waveforms. Accordingly, distance estimation using WLAN is generally possible from RSSI, ToA, TDoA and RTT.

8.4.1 Lateration Using RSSI

The RSSI value obtained by a wireless device is a function of the distance between emitter and receiver. The path loss or attenuation model (3.1) describes this dependency in general. However, this simplistic model does not fully describe the distribution of the path amplitudes in indoor environments, where the propagation conditions are of dynamic nature due to multipath (i.e. fast fading) and shadow fading (i.e. slow fading). Taking into consideration fast and slow fading the model becomes

$$P_R = P_T \frac{G_T G_R \gamma h^2}{4\pi d^p}, \qquad (8.4)$$

where P_R is the received power (RSS) value, P_T the transmitted power, G_T and G_R the transmitter and receiver antenna gains, d the distance, p the path loss exponent, γ a model parameter for slow fading (log-normal distribution) and h a parameter modeling the fast fading, often referred to as multipath fading. Fast fading appears in the form of small scale rapid amplitude fluctuations of the complex envelope, caused by reception of multiple copies of the signal through multipath propagation and is usually modeled by Rayleigh distribution in absence of a strong received component, i.e. NLoS conditions. In cases where a strong dominating path (e.g. the LoS path) exists compared to other arriving low level scattered paths, the Rician distribution is used. More details on the propagation models in indoor environments can be found in Hashemi (1993).

If the RSSI are averaged over a certain time interval, the fast fading term h in (8.4) can be approximated with $h = 1$ and using logarithmic units the dependency between RSSI P_R and distance d becomes

$$P_R = \alpha - 10\, p\, log_{10}(d) + z(0, \sigma), \qquad (8.5)$$

where the shadow fading (also known as slow fading) is modeled as suggested by Salo et al. (2007) as the zero-mean Gaussian random variable z with standard deviation σ. Shadow fading represents the slow variation in the signal amplitude due to obstacles in

8.4 WLAN Distance Based Methods (Pathloss-Based Positioning)

the propagation paths. The term α contains the averaged fast fading, the transmitter power P_T as well as the antenna gains G_T and G_R. Equation (8.5) can be used in any indoor environment to describe the relationship between RSSI values and the distance between an access point and a receiver. Figure 8.1 shows the logarithmic trend and also demonstrates that due to the logarithmic dependency, range estimation becomes more difficult with the increase in distance, i.e. long distances are almost indistinguishable from RSSI.

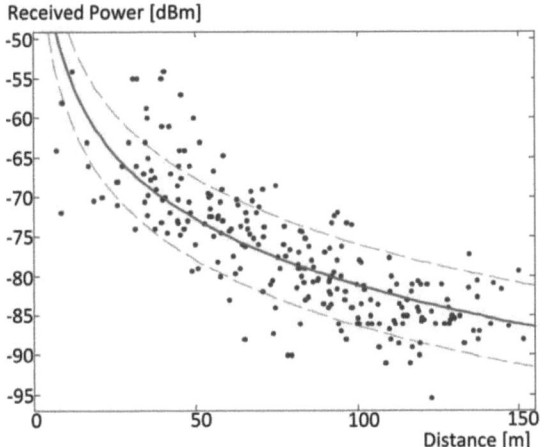

Figure 8.1 Dependency between distance and RSSI. The continuous line represents the log-distance model and the dots are measurements according to Laitinen (2004)

The main challenges for WLAN RSSI indoor positioning methods are the high time-variability of signal strength and the complexity of modeling the signal propagation according to attenuation patterns in indoor environments.

Mazuelas et al. (2009) use relation (8.5) to derive distance estimates from RSSI values to multiple WLAN access points in order to determine the location of a mobile receiver. To minimize the directional dependency of power transferred from transmitter to receiver caused by the rotational position of the antenna, omnidirectional antennas have been used. Prior to carrying out positioning, the averaged fading term α and the path loss p were determined for a specific indoor environment. Using an over-determined set of distance estimates for their multilateration approach, the redundancy was exploited to dynamically adjust the path loss exponents by least-squares minimization and also quantify the standard deviation of the determined position estimate based on the residuals. Mazuelas et al. achieved a mean positional accuracy of 4 m in their test measurements.

8.4.2 WLAN ToA

WLAN ToA based on trilateration constitutes a technology to overcome the offline training phase required by RSSI based techniques. Timing measurements also provide a more stable alternative compared to RSSI measurements with their known variability in time.

However, ToA observations are not directly available from a standard WLAN interface. The present WLAN standard does not provide timestamps with sufficient resolution in time. Assuming optimal conditions it is possible to access a time base of 1 µs using standard WLAN, which corresponds to a resolution of 300 m in distance. This problem might be mitigated in future with the introduction of IEEE 802.11v which has an allocation for timestamp differences within the request/response mechanism.

Methods using time delay measurements in standard WLAN are complex due to the difficulty of obtaining precise timing measurements. In order to precisely determine the impulse response, the distribution of the arrival time sequence has been studied by Hashemi (1993). Muthukrishnan et al. (2006) concluded in their feasibility study that ToA is not a feasible approach for localization for WLAN due to the limitation of current hardware and protocols.

For precise time measurements, it is preferred to keep the system operating at the lowest possible network layers in order to avoid extra delays due to processing between the layers. Therefore, most ToA ranging methods rely on precise time measurements at the physical layer at the cost of hardware modifications in the WLAN chipset. It should be noted however, that the use of modified WLAN hardware is an obstacle for deploying WLAN ToA as practical positioning solution.

The ToA approach of Golden and Bateman (2007) estimated distances between fixed access points and the location of a laptop. Their approach requires minor modifications to the physical layer (software and hardware). In order to mitigate multipath propagation three different measures are taken. First, a diversity scheme is applied which relies on a number of different communication channels. Secondly, the multipath is minimized by using directional antennas at the access points. Thirdly, the existence of different delayed paths in the received signal is taken into account by applying the path-decomposition-model (3.3). The reported accuracy for the distances is 1 m to 5 m.

Reddy and Chandra (2007) propose a ToA technique based on correlation of the WLAN signal where the received signal is correlated at the physical layer with training symbols stored in the receiver. Different channel models were defined which take into account specific power delay profiles typical of certain environments, e.g. an office with NLoS conditions. A major drawback for measurements at the physical layer is that specific (nonstandard) hardware modules are required, causing the implementation to be difficult to put through in real-world applications.

The estimation of ToA distances at upper layers of WLAN takes advantage of the WLAN (IEEE 802.11) standard protocol and therefore its applicability is facilitated for ranging using portable, off-the-shelf devices. Due to the limitations of unmodified WLAN hardware and protocols, current approaches determine ToA indirectly from Round Trip Time (RTT) ranging techniques.

8.4.3 WLAN TDoA

TDoA requires simultaneous reception of signals from at least 2 access points for the most basic operation mode. One premise of WLAN however is that only one node talks at a time, i.e. the access points do not transmit on the same channel simultaneously. Therefore, TDoA based on WLAN follows the other approach where the access points serve as multiple synchronized receivers and the mobile station transmits the signal. The problem is that neighboring access points are typically set to different frequency channels and listen only in that particular frequency.

8.4.4 WLAN AoA

IEEE 802.11n is an amendment to the WLAN standard to improve the network throughput by adding a technology which supports multiple antenna configurations, known as Multiple-Input Multiple-Output (MIMO). Wong et al. (2008) used this amendment of WLAN to determine the Angle of Arrival (AoA) of a mobile transmitter signal received at an access point with an array of four linear monopole antennas. From AoA of four access points, the position of the mobile was estimated. Based on simulations it was concluded that the potential positioning accuracy is better than 2 m.

8.4.5 WLAN RTT

The Round Trip Time (RTT) is the time difference between the time a pulse was sent to an access point and the time the response was received at the same device after return (see Section 3.3.3). In order to estimate a distance to a WLAN access point based on RTT, the delay at the access point must be known. WLAN access points have no deterministic delay. Due to variable interrupt latencies outgoing or incoming time stamps are falsified. The resulting delays have a typical variation of 5 μs which translates to 1500 m error in the distance estimation. One possible solution is to measure the time delay directly at the access point and forward the value of the delay to the mobile station. In addition, the clock drift of an RTT observation must be taken into account during transmission of a sequence. If a WLAN clock has a frequency stability of 50 ppm and the RTT observation takes 320 μs, an accuracy of 5 m can be assumed.

Günther and Hoene (2004) conducted experiments with WLAN RTT based on low-cost, commercial WLAN hardware (with modifications at the Media Access Control (MAC) layer but without modification at the physical layer). They used the intrinsic feature of WLAN that a ping response is immediately acknowledged by its receiver to determine indirectly the air propagation time. By using multiple delay observations and applying statistical methods (i.e. stochastic resonance) Günther and Hoene found that the propagation delays correlate closely with the distance, obtaining a deviation of only a few meters.

Ciurana et al. (2009) carried out a pure software based solution for two-way ranging on the IEEE 802.11 standard protocol and off-the-shelf WLAN hardware. From the Round Trip Time RTT_i for a distance i and the reference Round Trip Time RTT_0 for the distance zero, the signal travel time

8 WLAN / Wi-Fi

$$\Delta t_i = \frac{RTT_i - RTT_0}{2} \tag{8.6}$$

can be determined. The RTT is the difference between time-stamps obtained at the MAC layer. Experiments conducted under LoS conditions showed a large dispersion of RTT values. After averaging 1000 measurements (equivalent to a duration of 1 s), a ranging accuracy of 1.7 m was obtained for a LoS indoor distance of 13 m. An attempt to reduce the dispersion of the RTT observations has been made by Ciurana et al. (2010). In order to become a practical solution for indoor positioning, more hostile scenarios must be addressed and the impact of simultaneously injected data traffic at the access point must be better understood. Based on simulation, Tappero et al. (2010) predict that a campus wide implementation of a WLAN RTT positioning system can achieve an accuracy of 3 m to 5 m.

8.5 Summary on WLAN Systems

Fingerprinting based on RSSI values is the prevalent method of using WLAN for positioning. Depending on the density of calibration points, fingerprinting reaches accuracies of 2 m to 50 m, see Table 8.1. The fingerprinting method is particularly of commercial interest, because off-the-shelf devices can be used. Experiments on WLAN time-of-arrival distance measurements have proven to be of poor quality due to multipath and low resolution of the clocks. Using RSSI for distance estimation in indoor environments has also proved unreliable due to an irregular dependency between attenuation and distance in indoor environments.

Table 8.1 Localization approaches using WLAN and reported performance parameters

Name	Year	Reported Accuracy (m)	Area (m²) per Access Point. For non RSS: Type of Modification	Principle	Prior calibration	Number of Calibration Points	Model / Method	Market Maturity
Parodi	2006	3.3	1235 / 14 = 88	online learning	minimal	114	DPM	study
Xiang	2004	2 – 5	1400 / 5 = 280	fingerprint & map	yes	100	offline training	study
Bahl	2000	5	978 / 3 = 326	fingerprinting	yes	70	offline training	study
Gansemer	2010	2.1	972 / 24 = 40	fingerprinting	yes	972	offline training	study
Koski	2010	5 – 7	5600 / 206 = 27	fingerprint & map	yes	96	coverage area	study
Hansen	2010	2	ca. 140/ 14 = 10	fingerprinting	yes	17	dynamic model	study
Chen	2005	2 – 4	400 / 5 = 80	fingerprint & RFID	no	online	dynamic model	study
King	2006	1.6	312 / 20 = 156	fingerprinting	yes	166	offline training	study
Teuber	2006	2 – 3	400 / 5 = 80	fingerprinting	yes	16	fuzzy logic	study
Mazuelas	2009	4	3375 / 8 = 422	multilateration	no	only AP	path loss model	study
Gallagher	2010	room level	unknown	fingerprinting	yes	low	signal distance	study
Ekahau	2009	7	ca. 450 / 5 = 90	fingerprinting	yes	-	offline training	product
Skyhook	2011	30 – 70	'global' coverage	fingerprinting	no, by Skyhook		offline training	product
Golden	2007	1 – 5	hardware mod.	diversity	no	-	path decomposition	study
Reddy	2007	15	hardware mod.	time of arrival	no	-	LTS correlation	simulation
Wong	2008	2	hardware mod.	angle of arrival	no	-	-	simulation
Günther	2004	5-15	software mod.	round trip time	no	-	-	study
Ciurana	2009	1.7	software mod.	round trip time	no	-	line of sight	study

9 Radio Frequency Identification

Wavelength	100 nm	1 μm	10 μm	100 μm	1 mm	10 mm	0.1 m	1 m	10 m	100 m	1 km
	UV		Infrared								
Frequency	3 PHz	300 THz	30 THz	3 THz	300 GHz	30 GHz	3 GHz	300 MHz	30 MHz	3 MHz	300 kHz

An RFID (Radio Frequency IDentification) system consists of a reader with an antenna which interrogates nearby active transceivers or passive tags. Using RFID technology, data can be transmitted from the RFID tags to the reader (also known as RFID scanner) via radio waves. Typically, the data consist of the tag's unique ID (i.e. its serial number) which can be related to available position information of the RFID tag. The most frequently employed positioning principle is that of proximity, also known as CoO (Cell of Origin), e.g. the system indicates the presence of a person wearing an RFID tag. Thereby, the accuracy of an RFID system is highly depending on the density of tag deployment and the maximal reading ranges. Alternatively, the Received Signal Strength Indicators (RSSI) can be used for coarse range estimation in order to apply multilateration techniques. Time of Arrival (ToA) distance estimation on RFID has been proven difficult to achieve. In order to measure the distance between a reader and a tag at a resolution of better than one meter, a bandwidth of at least 10 kHz must be used and multiple observations need to be averaged. The standard case of RFID positioning based on ToA distances relies on observations from a single tag, where the position determination is carried out in combination with Angle of Arrival (AoA) measurements from that tag. Also Phase of Arrival (PoA) methods have been proposed for RFID, see Povalač and Šebesta (2010). Fingerprinting (FP) based on pre-measured signal maps can also be applied. Readers are able to scan several tags at high data rates up to 10 Hz.

Generally, RFID systems can be unobtrusive to the user by integrating the tags in the pavement, under the carpet or in the walls without direct line of sight, since the radio waves have the ability to penetrate solid materials to some extent. The higher the frequency, the more the signal suffers from attenuation. The typical frequency ranges are categorized as Low Frequencies (LF) at 30 kHz to 500 kHz, High Frequencies (HF) at 3 MHz to 30 MHz, Ultra-High frequencies (UHF) at 433 MHz & 868 MHz to 930 MHz and microwaves (SHF) at 2.4 GHz to 2.5 GHz & 5.8 GHz.

Products equipped with RFID tags can be tracked seamlessly from a supplier's factory to the retailer's store shelves. However, as RFID moves to item and device tagging, the probability of tags disclosing personally identifiable information has become a primary privacy concern for users. On the other hand, the unique identification of RFID can be used for security to control building access or for payment systems allowing customers to pay for items automatically. When scaling an RFID system to large numbers in multiple facilities with geographical distribution, a challenge for data management and configuration arises.

9.1 Active RFID

An active RFID system consists of deployed RFID scanners which interrogate active radio transceivers equipped with internal battery power. In contrast to passive tags the need for batteries makes active transceivers heavier and more costly, but enables long detection ranges of 30 m or more. The active RFID technology can be used for positioning, where the location estimation is typically carried out by fingerprinting on RSSI.

Seco et al. (2010) have achieved a median positioning accuracy of 1.5 m based on 71 active RFID tags covering 55 rooms of 1600 m^2. Using Gaussian processes to describe the spatial dependence of RSSI signals propagating indoors they improved the accuracy by 30% versus least squares minimization. Jiménez et al. (2010) used the active RFID tags to stabilize the IMU drift of a foot mounted pedestrian navigation system. Their method of tight IMU-RFID integration limits the absolute drift deviation to maximal 1 m to 3 m. At an emission frequency of 1 Hz, it is expected that the tags' batteries will last for a period of 6 months.

Peng et al. (2011) integrated an active RFID system with INS/GNSS in order to establish a seamless indoor/outdoor positioning system for the purpose of vehicle and pedestrian navigation. Based on a Sigma Point Kalman Filter multi-sensor integration algorithm, meter-level accuracy could be achieved.

Kimaldi (2011) offers a solution for applications in hospitals based on static active readers operating in the microwave band. The tags are worn as wristbands or attached to keyrings for access control and personal monitoring.

9.2 Passive RFID

Passive RFID systems solely rely on inductive coupling and therefore don't require batteries. The principle of inductive coupling allows the tags to receive sufficient energy in the form of RF waves from the nearby RFID scanner to transmit their codes back to the scanner. Passive tags can be applied for waypoint navigation based on a reference grid of ID markers whose locations are accessible from a database. The advantages of using passive RFID tags for positioning are their small size, high level of ruggedness, relatively inexpensive installation and low maintenance needs since they have no batteries. Therefore, passive tags are suitable for subsurface embedding in building material. As the main drawback, the detection range is usually limited to 2 m, which demands dense deployment of tags. The attenuation for embedded tags increases with the frequency, e.g. the application for concrete embedded tags is prohibitive for frequencies above 2.4 GHz. Below 100 MHz the attenuation of embedded tags is low and comparable to that of free space, but frequencies below 300 MHz are unsuited for read ranges above 1 meter due to inductive coupling as a phenomenon of near-field operation. The penetration of electromagnetic waves in concrete depends on the use of integrated metal fibers, the moisture content of walls, floors and ceilings and the angle of incidence of the radio wave. Pena et al. (2003) have shown that signal propagation through reinforced concrete has a power loss of 90% at 30 cm penetration depth and a frequency of 900 MHz.

Daly et al. (2010) have embedded passive RFID tags in concrete and optimized the antenna design with respect to a maximal reading range for passive tags. They were able to double the readability by achieving a reading range of up to 1.2 m of concrete embedded tags. The embedding of RFID in the ground is used for navigation based on smart paving stones.

Further applications are discovery and identification of buried pipes. Dziadak et al. (2005) suggest attaching passive RFID tags to buried non-metallic pipes in order to facilitate their detection.

Passive tags have shown to be applicable for vehicle guidance (Baum et al. 2007) as well as inventory control (Frick 2011).

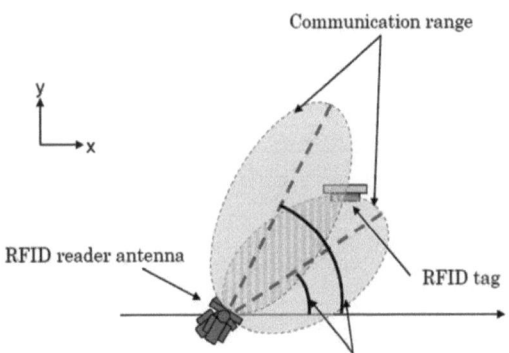

Figure 9.1 Intersection of two RFID antenna communication boundaries based on different rotation angles according to Uchitomi et al. (2010)

The navigation system 'ways4all', developed by Kiers et al (2011) includes arrays of passive RFID tags which have been deployed under the carpet to provide route guidance for visually impaired and blind people. Experiments show, that the current maximal reading range of 30 cm needs to be extended to 60 cm in order to obtain a sufficiently high tag detection rate.

9 Radio Frequency Identification

Uchitomi et al. (2010) have enhanced the pure proximity positioning of an RFID reader antenna based on passive RFID tags. They rotate the RFID reader antenna in the horizontal plane and estimate its position from the intersection of positions of two communication boundary areas, see Figure 9.1. Using this method, a robot arm of 1.2 m in length could be located with an accuracy of less than 20 cm.

Another enhancement of passive RFID positioning based on proximity detection has been implemented by Fujimoto et al. (2011). By varying the transmission power of the reader, they create different sensing ranges to refine the distance estimation of a tag. In addition, the reader is moved to different locations in order to determine its position with higher accuracy. It takes 30 seconds to determine a position.

Future Shape offers the solution for robot navigation based on a 50 cm grid of passive RFID tags with the product name NaviFloor (2011). One application is that a cleaning robot can record the date and place of its operation and even which cleanser was used.

9.3 Summary on RFID Systems

Most RFID systems rely on proximity detection of permanently mounted tags to locate mobile readers. Therefore the accuracy of an RFID system is directly related to the density of tag deployment and reading ranges. Some long-range active RFID systems can also use signal strength information to improve the localization accuracy. RFID can be combined with IMU in order to correct the drift which IMUs have. The main application of RFID location systems is route guidance for pedestrians. System parameters of RFID approaches included in this study are given in Table 9.1

Table 9.1 Localization systems using RFID and reported performance parameters

Name	Year	Active Passive	Tag Range	Tag Frequency	Tag Deployment	Accuracy	Positioning Principle	Application	Market Maturity
Seco	2010	active	30 m	433 MHz	walls	1.5 m	RSSI, FP	person/object location	development
Jiménez	2010	active	70 m	433 MHz	walls	1-3 m	RSSI + IMU	pedestrian navigation	development
Peng	2011	active	100 m	915 MHz	floor	1-3 m	RSSI + IMU	pedestrian navigation	study
Kimaldi	2011	active	13 m	2.45 GHz	wrist, keyring	room	CoO	hospital	product
Kiers	2011	passive	11-30 cm	134 kHz	under carpet	dm	CoO	navigation of blind	prototype
Daly	2011	passive	1.2 m	868 MHz	concrete	m	CoO	vehicles and blind	prototype
Dziadak	2005	passive	2 m	130 kHz	soil, 2m depth	m	CoO	buried asset detection	study
Baum	2007	passive	7-10 cm	13.5 MHz	foil cover road	dm	CoO	guided vehicle	development
Utchitomi	2010	passive	2 m	2.45 GHz	floor	20 cm	CoO + AoA	pedestrian navigation	simulation
Fujimoto	2011	passive	2 m	2.45 GHz	floor	15 cm	CoO + range	pedestrian navigation	study
NaviFloor	2011	passive	> 50 cm	13.5 MHz	floor	50 cm	CoO	robot navigation	product

10 Ultra-Wideband

Wavelength	100 nm	1 μm	10 μm	100 μm	1 mm	10 mm	0.1 m	1 m	10 m	100 m	1 km
	UV		Infrared				Microwave		Radio		
Frequency	3 PHz	300 THz	30 THz	3 THz	300 GHz	30 GHz	3 GHz	300 MHz	30 MHz	3 MHz	300 kHz

Ultra-Wideband (UWB) is a radio technology for short-range, high-bandwidth communication holding the properties of strong multipath resistance and to some extent penetrability for building material which can be favorable for indoor distance estimation, localization and tracking. A typical UWB setup features a stimulus radio wave generator and receivers which capture the propagated and scattered waves. In contrast to narrowband operation, UWB waves occupy a large frequency bandwidth (>500 MHz). More precisely, an emitted radio wave belongs to UWB if either the bandwidth exceeds 500 MHz or 20 % of the carrier frequency. In order to avoid interference with other radio services, the Federal Communications Commission (FCC) in the USA has limited the unlicensed use of UWB to an equivalent isotropically radiated power density of -41.3 dBm/MHz and restricted the frequency band to 3.1 GHz - 10.6 GHz (respectively 6.0 GHz - 8.5 GHz in accordance to the European Communications Committee (ECC)). Figure 10.1 illustrates how UWB coexists with other Radio Frequency (RF) standards.

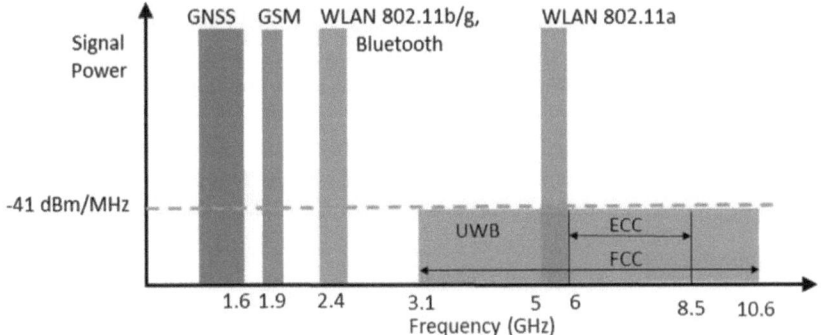

Figure 10.1 Regulated UWB spectrum

Legal restrictions in signal power limit the operating range to less than 100 m. On the other hand the low power spectral density prevents harmfulness to the human body and bounds the interference of UWB signals with other narrowband receivers. Licensed UWB technologies operate in the wavelength of microwaves, where the low frequency components in the UWB signal spectrum have the ability to penetrate building materials such as concrete, glass and wood (Koncur et al. 2009). This is a useful property for indoor positioning, because it enables ranging under NLoS conditions and makes inter-room ranging possible. On the other hand, partial signal penetration into the target object is unfavorable for precise distant measurements, because the reflected signal includes multiple returns besides the outer boundary reflection. Therefore robust extraction of useful information from the received signal is a major challenge in UWB ranging. Superposition of different scattering effects complicates data interpretation.

10.1 Range Estimation Using UWB

A major advantage of using UWB for distance measurements is that large bandwidth translates into a high resolution in time and consequently in range. The achievable range resolution rr can be approximated with

$$rr \approx \frac{v}{2b}, \qquad (10.1)$$

where v is the speed of the wave front and b the bandwidth. E.g. for the FCC band and propagation in free space (assuming speed of light with $v = c_0$) it is $rr \approx 0.5\ c_0\ /\ 7.5\ \text{GHz} = 2\ \text{cm}$, respectively 6 cm for the ECC band at bandwidth $b = 2.5\ \text{GHz}$).

UWB ranging techniques include Time of Arrival (ToA), Two Way Ranging (TWR), Time Difference of Arrival (TDoA). All these techniques rely on time measurements which can be divided into three different principles as described below.

10.1.1 Continuous Waves

Within the frequency band, different frequencies are sequentially used by stepping or sweeping (i.e. frequency modulation in a manner similar to a chirp, see Figure 13.2). The signal is analyzed in the frequency domain resulting in low time resolution which is unfavorable for dynamic real-time applications. Continuous waves allow for precise ranging, but cannot be used for small devices such as a smart phone because such technology requires large antennas. If the frequency range is very wide, a large physical size of the antenna is necessary to achieve sufficient antenna efficiency.

10.1.2 Impulse Radio

The UWB Impulse Radio (UWB-IR) is simply structured and can be used for fast distance measurements. The duration of the pulses is in the order of nanoseconds or even less. Compared to continuous waves, ultra-short pulses are less likely to interfere with signals traveling other paths allowing for better resolution of the line of sight path and therefore evoking robustness against multipath. Since the radios have to be powered for a short time only before and during pulse generation, UWB-IR has a low power consumption compared to other UWB techniques. UWB pulse-based ranging systems are employed with relatively low repetition rates of 1 MHz to 100 MHz (i.e. megapulses per second), in contrast to UWB communication systems which use 1 GHz to 100 GHz (Robert et al. 2010). Stoica (2006) presents a demonstrator for UWB-IR ranging based on non-coherent energy-collection and achieves 1.5 m (5 ns) distance estimation. Fischer et al. (2010) use impulses of 300 ps duration (equivalent to 9 cm wavelength) employing a Two Way Ranging (TWR) technique based on the gated oscillator principle. An accuracy of 4 cm was reported under perfect LoS conditions. Pietrzyk and von der Grün (2010) achieved a ranging accuracy of 1 cm to 2 cm for LoS distances of up to 5 m between an UWB-IR transmitter and a non-coherent energy detection receiver.

An actual implementation for localizing a mobile robot in industrial environments with NLoS conditions has been carried out by Segura et al. (2010). Four synchronized anchor nodes are used to determine the 2D location of a robot within 20 cm accuracy by TDoA ranging and hyperbolic intersection. The method also works under NLoS conditions at slightly poorer positional accuracy.

10.1.3 Pseudo Noise Modulation

Random or pseudo noise can be applied for ToA ranging (Herman et al. 2010, Zetik et al. 2010, Kocur et al. 2009). Usually this technique is based on Maximum Length Sequences (MLS, also known as M-sequences) which are pseudo-random binary sequences (of 585 ns duration for example). As a disadvantage, large processing capabilities are required for determining the correlation at the receiver. Nevertheless, pseudo noise techniques are of interest for the mobile phone market, because miniaturized antennas can be used.

10.2 Multipath Mitigation Using UWB

The large signal bandwidth of UWB allows for high resolution in time such that a large number of delayed multipath components appear in the signal and become resolvable. The rich multipath diversity can be used to mitigate signal fading. According to Hausmair et al. (2010) it is an ongoing research challenge to extract the multipath components from the information embedded in the Channel Impulse Response (CIR) of the UWB signal. Molisch (2009) has made an attempt to better understand the UWB fading statistics and CIR which consists of the sum of delayed, attenuated and scaled replicas of a transmitted impulse.

Coherent receivers are particularly well suited for multipath mitigation. A coherent receiver recovers the absolute phase, amplitude and time of the received multipath components, while a non-coherent receiver uses only the envelope of the signal power by autocorrelation of the received signal. Since non-coherent receivers do not use frequency and phase, less information is provided for the separation of multi-path reflections. Therefore, their ability to detect multipath is limited compared to coherent receivers.

10.3 Positioning Methods Using UWB

10.3.1 Passive UWB Localization

Passive UWB systems have the potential for localization of people and objects from signal reflection, using the principle of radar (Chapter 13.5). The advantage of passive UWB is that neither an active nor a passive tag needs to be worn by the user nor attached to the localized object. A typical passive localization setup includes one or more omnidirectional emitter antennas and multiple listener antennas, as shown in Figure 10.2.

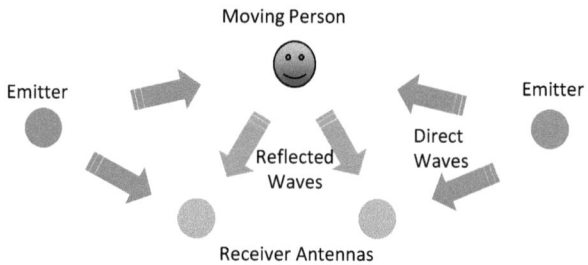

Figure 10.2 Passive localization setup

In order to detect a moving person, time-invariant direct waves and strong static signals reflected from furniture need to be subtracted. Given that the locations of the emitter and receiver antennas are known, estimated ranges from the reflected waves can be used for any range-based algorithm such as ToA or TDoA multilateration to determine the object position. Following the so-called background subtraction approach, it is possible to observe extremely small movements such as cardiac or respiratory activity (e.g. Kosch et al. 2011).

Based on experiments with a network of deployed UWB pseudo-noise sensors, Herrmann et al. (2010) have shown the feasibility of UWB for application in Ambient Assistant Living (AAL). The test network consisted of two transmitters and four receivers in one room. From signal reflections from a moving person, the propagation distance and the reflectivity of the received wave were used to determine the location and identify the activity of the person. The authors concluded that scattering effects at the human body complicate data interpretation. For the purpose of smart audio systems, Zetik et al. (2010) demonstrated that passive UWB localization has the capability to detect walking persons by employing a setup of multi-receiver antennas.

10.3.2 UWB Virtual Anchors

Under the assumption that the 3D geometry of a room is known, multipath signal reflections from a single UWB transmitter can be used for position estimation of a mobile receiver. Even if only one transmitter has been physically deployed, signal reflections from walls cause path delays which can be used to derive pseudodistances to

10.3 Positioning Methods Using UWB

so-called virtual anchors and perform multilateration to estimate the transmitter position. Figure 10.3 illustrates the principle of virtual anchors. Meissner et al. (2010) propose the virtual anchor method in combination with the use of trajectory information. To determine the unknown positions, a multimodal optimization problem is solved by a so-called state space estimation technique.

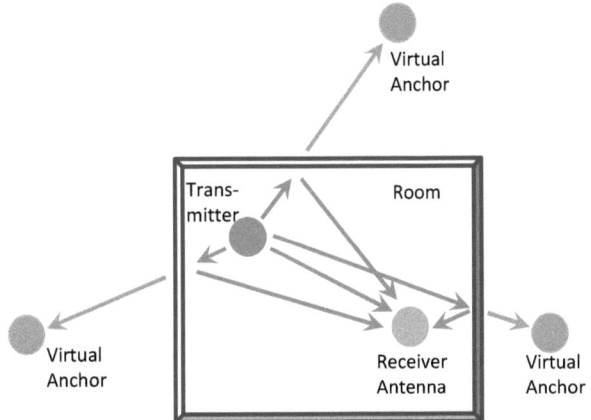

Figure 10.3 Virtual anchor based multilateration

10.3.3 UWB Direct Ranging

Direct, non-reflected UWB range measurements can also be used for positioning. As active tags are carried by the mobile user, such approaches can be categorized as active systems. Implementations can be based on ToA, TDoA or RTT (Round Trip Time) distance measurements from UWB Impulse Radios (UWB-IR, see 10.1.2.). Positions are determined based on LoS distances by lateration techniques.

10.3.4 UWB Fingerprinting

The large frequency spectrum of a Channel Impulse Response (CIR) from an UWB transmitter contains a large amount of information originating from multiple multipath components. Therefore, a unique fingerprint can be assigned to each location and stored in a database during an offline training phase. In an online phase, the current CIR is compared to the database and the fingerprint which is responsible for the optimal cross correlation coefficient is returned as the current position. UWB fingerprinting has the potential to achieve better accuracies than fingerprinting based on WLAN and is also applicable in NLoS scenarios. The advantage of fingerprinting using Received Signal Strength Indicators (RSSI) is that the method does not require time synchronization. As a drawback, the RSSI method does not make use of the full potential of high resolution available with UWB.

Kröll and Steiner (2010) implemented UWB fingerprinting with a grid spacing of 1 cm. Their demonstrator system consisted of one transmitter, one receiver and 6 fingerprinting regions, each 28 cm × 28 cm in size. The research team was able to correctly identify each neighbor region and achieved an average position accuracy of 4 cm.

Wang et al. (2010) carried out fingerprinting based on RSSI to localize a static UWB receiver. Based on a demonstrator system setup, a positional accuracy of 0.33 m was reported after 20 fingerprints had been taken from four fixed UWB transmitters on an area of 1.2 m × 1.6 m.

10.4 Commercial UWB Systems

A commercially available UWB-IR positioning solution is that of Ubisense (2011). The system operates in the 6.5 GHz to 8 GHz band and localizes small sized active tags equipped with long-life batteries. A system setup requires deployment of multiple fixed receivers which include antenna arrays for the estimation of the angle of incidence. Position determination is carried out by a combination of TDoA and AoA with a specified 3D accuracy of 15 cm for indoor environments and an operating range of around 50 m.

The UWB-IR system of Zebra Enterprise Solutions (2011) utilizes TDoA to locate tags up to a distance of 100 m and has a specified accuracy of better than 30 cm under line of site conditions. The system is particularly well suited for environments with severe multipath.

10.5 Summary on Ultra-Wideband Systems

Since achievable accuracy of ToA measurements is directly correlated to the signal bandwidth, Ultra-Wideband is well suited for precise ranging. At bandwidths of several hundred MHz and appropriate time resolution in the order of nanoseconds, ranging and positioning at cm-level are possible.

Table 10.1. Localization systems using UWB and performance parameters as reported.

Name	Year	active / passive	Transmitters	Receivers	Noise Radar or IR (Pulse Duration)	Principle	Application	Accuracy	Market Maturity
Robert	2010	active	1	4	IR (1.5 ns)	ToA	indoor application	not reported	demo.
Stoica	2006	active	1	1	IR (750 ps)	ToA	sensor networks	1.5 m	demo.
Fischer	2010	active	1	1	IR (300 ps)	ToA, RTT	industrial application	4 cm	demo.
Pietrzyk	2010	active	1	1	IR (65 ps)	ToA	precise ranging	1 cm – 2 cm	demo.
Segura	2010	active	4	1	IR (2 ns)	TDoA	mobile robot	20 cm	prototype
Herrmann	2010	passive	2	4	noise M-seq.	body reflection	AAL, monitoring	person detection	demo.
Zetik	2010	passive	1	4	noise M-seq.	body reflect., ToF	smart audio system	person detection	demo.
Kocur	2009	passive	1	2	noise M-seq.	reflect, ToA	through wall pos.	person detection	demo.
Meissner	2010	passive	1	1	IR	virtual anchors	tracking	0.5 m	simulation
Kröll	2010	both	1	1	pseudo noise	fingerprinting	office	4 cm	demo.
Wang	2010	active	4	1	-	RSSI fingerprint	-	0.3 m	demo.
UBISENSE	2011	active	>99	>2	IR (very short)	TDoA, AoA	automation	< 15 cm	product
Zebra	2011	active	>99	>2	IR (short)	TDoA	logistics	< 30 cm	product

UWB is especially useful for indoor environments where multipath is severe, because the wide bandwidth facilitates the detection of multiple time-delayed versions of a signal sequence. The reason why UWB has not entered the mass market (apart from a few industrial implementations) is the requirement for a dedicated transmitter-receiver infrastructure. An overview of UWB implementations included in this study is given in Table 10.1.

11 High Sensitive GNSS / Assisted GNSS

Wavelength	100 nm	1 µm	10 µm	100 µm	1 mm	10 mm	0.1 m	1 m	10 m	100 m	1 km
	UV		Infrared			Microwave			Radio		
Frequency	3 PHz	300 THz	30 THz	3 THz	300 GHz	30 GHz	3 GHz	300 MHz	30 MHz	3 MHz	300 kHz

Employing Global Navigation Satellite Systems (GNSS) for positioning has the potential to fulfill the vision of ubiquitous positioning. GNSS are the only technique which satisfies the two conditions: a) independent operation from local infrastructure and b) world-wide coverage. However, indoor spaces are hardly covered by GNSS because receivers are challenged with the task of tracking satellites in an environment where signal fading is severe. The difficulty of receiving satellite signals indoors has triggered promising developments involving the building of highly sensitive and assisted GNSS (Global Navigation Satellite Systems) receivers – with many issues remaining unsolved.

11.1 Signal Attenuation

The power of the C/A code of the L1 frequency at the GPS satellite is about 500 Watts (27 dBW) if the antenna gain is taken into account. The free space path loss as a function of the satellite-to-receiver distance of 21'000 km is about 185 dBW (18 orders of magnitude). Therefore, a signal power of 27 dBW − 185 dBW = -158 dBW = -128 dBm can be expected at the Earth surface. The relevant quantity for signal acquisition is the receiver specific signal-to-noise-ratio (SNR) which depends not only on the received signal power but also on the type of receiver amplifier.

Apart from free space path loss, building material causes additional attenuation of the satellite radio signal in indoor environments, where the attenuation factor depends primarily on electrical properties of the material such as the dielectric coefficient. Penetration of GNSS signals through ferro-concrete walls causes attenuation of 20 dB to 30 dB (a factor of 100 to 1000) compared to the signal amplitudes outdoors. Loss factors for various materials are listed in Table 11.1. While attenuation in wooden residential homes is moderate with 5 dB to 15 dB, typical brick or concrete buildings cause a loss of 20 dB to 30 dB. In underground car parks and tunnels the already weak satellite signals become nearly undetectable for GNSS receivers at loss factors of >30 dB, see Table 11.2.

Greater transmitter power would improve the signal strength and lighten the impact on attenuation but can hardly put into practice with satellites which solely rely on solar panels for their power.

Table 11.1 Attenuation of various building materials for the L-Band (L1 = 1500 MHz), Stone (1997)

Material	(dBW)	Attenuation Factor ( ) for a Typical Thickness
dry wall	1	0,8
plywood	1 – 3	0,8 – 0,5
glass	1 – 4	0,8 – 0,4
painted glass	10	0,1
wood	2 – 9	0,6 – 0,1
iron mat	2 – 11	0,6 – 0,08
roofing tiles / bricks	5 – 31	0,3 – 0,001
concrete	12 – 43	0,06 – 0,00005
ferro-concrete	29 – 33	0,001 – 0,0005

Table 11.2 Signal strengths in decibel watt (decibels relative to one watt)

Environment	Signal Strength (dBW)	Difference to Outdoors (dB)	Comment
satellite	+ 27	+185	reference signal strength delivered from satellite
outdoors	-158	0	nominal carrier power received at receiver (Joseph 2010)
indoors I	-176	-18	indoor environments near windows, urban canyons
indoors II	-185	-27	inside office buildings, multilevel car parks
underground	-191	-33	decode limit for aided, ultra-high sensitive receivers

Apart from attenuation of the GNSS signal, the topic of indoor signal reception is more complex and high sensitivity is only one milestone. Phenomena such as reflections, diffraction or scattering occur when radio signals enter a building. In particular, multipath reception induces individual time shifted measurements which degrade the quality of the received GNSS signal and lead to less reliable positioning. Indoor fading caused by building material is less well understood compared to the fading taking place in the troposphere and ionosphere which can be described by existing models. See Hein et al. (2008) for more details.

11.2 Assisted GNSS

Assisted GNSS (AGNSS, AGPS) is a thoroughly standardized positioning method in outdoor environments. Applications of AGNSS include localization of mobile phones or devices with internet access. AGNSS receivers employ an additional data link to provide information of satellite ephemeris, almanac, differential corrections and timing information which is normally obtained from the GNSS satellites directly.

11 High Sensitive GNSS / Assisted GNSS

A cold start (i.e. initial time, position and almanac being inaccurate) of an unassisted GNSS receiver requires search through a 2D space of about 30 frequency bins and 1000 code-delay chips (i.e. 30 × 1'023). The search through 30 frequency bins is necessary because the relative motion between receiver and satellites causes a shift in frequency (i.e. Doppler-shift). Since the Doppler shift is unknown in case of a cold start, the receiver must search across a wide frequency range, typically 30 Doppler bins. Since it takes about 1 s to search all code-delays of one frequency, the correlation peaks for all frequencies can be found within 30 s. It takes another 30 s until the time-of-week and ephemeris data is decoded, such that the total Time To First Fix (TTFF) is approximately 60 s. In a weak signal environment, the first fix takes longer or is even not possible at all, because any lost data bit requires the receiver to obtain another data message of 30 s. In case of AGNSS, additional data helps to reduce the TTFF significantly in two ways. First, it is possible to save the timespan of 30 s which an unassisted receiver spends on downloading the ephemeris data from the satellites. Secondly, the search for the correlation peak is facilitated, because an AGNSS enabled receiver has prior information of each satellite's position and velocity from an AGNSS server which can be used to effectively reduce the number of frequencies to be searched. By providing time, frequency and ephemeris data via the additional data link, a fix position is obtained much faster – under the assumption that at least some weak signals to four satellites can be received. An AGNSS receiver has therefore an increased sensitivity, facilitating the use of signal strengths below the normal threshold to carry out pseudorange measurements. Therefore AGNSS is an enabler for position estimation with GNSS in various indoor environments. Details on technical aspects of AGNSS can be found in Diggelen (2009).

11.3 Long Integration and Parallel Correlation

Apart from AGNSS, High Sensitive GNSS (HSGNSS) receivers make use of two further techniques which facilitate reception of weak satellite signals indoors. The first technique improves low signal to noise ratio by integration over multiple intervals, while at the same time accepting longer acquisition times. Conventional GNSS receivers integrate received signals for 1 ms, which is the duration of a C/A code cycle. The integration time can be increased up to 20 ms or even longer by predicting bit transition in the navigation message. However, this method can lead to unacceptably long searching durations. The second technique relies on massive parallel correlators to reduce the computing time for correlation by a factor of 500 or more and lessen the required receiver power (Eissfeller et al. 2005). AGNSS in combination with massive parallel correlation (e.g. 10^5 correlators) can still not satisfy most applications in deep indoor scenarios. Within the next 10 years, a new level of indoor performance with GNSS might be reached with an improved GPS signal structure, GLONASS and additional use of Beidou and Galileo signals.

Currently, with the use of 44'000 correlators and SBAS (Satellite Based Augmentation System) 10 m accuracies can be achieved in indoor environments, e.g. Opus III (eRide 2011). Skyhook (2011) reports an accuracy of 10 m with an availability of 99.8 % and a Time-To-First-Fix of 4 s for its hybrid XPS system which combines WLAN, cell tower and AGNSS positioning. Lachapelle (2004) achieved an accuracy of 58 m over integration intervals of 10 s to 20 s in a commercial building and a horizontal positioning accuracy in the order of 10 m in a residential home.

Wieser (2006) assessed performances of a low-cost high sensitive u-blox antenna in partially obstructed scenarios. A high sensitive receiver was able to observe a significantly higher number of pseudo ranges to the satellites compared to a standard receiver, but the pseudo-range accuracy was 1 to 2 magnitudes lower compared to unobstructed environments (70 m instead of 4 m). Wieser showed that, taking into account a proper variance model and quality control, the positioning accuracy in harsh environments with the majority of observations obtained from obstructed satellites could be improved significantly (down to 5 m to 20 m).

Zhang et al. (2011) have assessed four static AGNSS enabled HSGNSS receivers in various indoor environments and obtained accuracies of 20 m to 60 m, see Table 11.3. Their overall conclusion is that there is no clear 'winner' on the HSGNSS receiver market today.

To improve the understanding of GNSS indoor reception, Kjærgaard et al. (2010) compared performances of HSGNSS receivers, including those embedded in mobile phones. The result of an extensive measurement campaign was that position accuracy of 5 m can be achieved inside wooden houses, 10 m in most concrete buildings and around 20 m in shopping malls. The significant factors degrading GNSS performance are the number of overlaying floors and the building material. GNSS receivers embedded in mobile phones generally provided slightly lower performances.

Eissfeller et al. (2005) show that acquisition of GNSS signals in environments facing severe signal attenuation of 25 dBW or more (e.g. basements, concrete buildings) requires Assisted GNSS. According to Eissfeller et al., currently achieved accuracies in office buildings of 20 m don't allow room identification, but this could be realized with differential corrections using DGNSS.

Soloviev and Dickman (2010) extracted GNSS carrier phase measurements 30 dBW below open-sky conditions (deep indoors), achieving cm-level accuracies. The approach uses signal integration times of 1 s and advanced processing techniques to track the carrier phase. A precondition for coherent integration of 1 s is the generation of a long and precise signal replica which requires short-term clock stability of 10^{-10} s/s and reproduction of user motion, i.e. the inertial drift must not exceed 1 cm/s which can be achieved by employment of a commercial-grade IMU.

11.4 Summary on High Sensitive GNSS

Performance of indoor GNSS using high sensitivity technologies has been shown to be severely degraded, as compared to the level of performance achievable in outdoor environments. GNSS can be used inside buildings made of wood or bricks at accuracies in the order of 10 m when accepting acquisition times around 20 s. However, HSGNSS is not yet ready to be used for pedestrian navigation in most public buildings and therefore the market of emerging location-based services cannot be served with satisfaction based on HSGNSS alone. However, HSGNSS can be a useful component of an IMU multi-sensor fused indoor navigation system to provide sparse position updates. Table 11.1 gives an overview of various HSGNSS solutions and experimental studies.

Table 11.3 Performance of AGNSS enabled high sensitive receivers during static tests

Name	Year	Indoor Environment	Strategy	Receiver	Horizontal Accuracy (m)	Vertical Accuracy (m)	TTFF (s)
eRide	2011	typical -191dBW	massive parallel	OPUS III	10.0	-	20.0
Skyhook	2011	difficult environ.	plus WLAN	SiRF	10.0	-	4.0
Lachapelle	2004	garage	massive parallel	SiRF-Star II	10.0	-	hot start
Lachapelle	2004	office building	massive parallel	SiRF-Star II	58.0	-	10 – 20
Wieser	2006	harsh environm.	variance model	u-blox TIM-LH	7.7	8.4	10.0
Zhang	2011	lecture room	AGNSS	Navman Jupiter 32	18.4	26.4	24.8
Zhang	2011	lecture room	AGNSS	SiRF GSCI-5000	53.4	61.2	11.8
Zhang	2011	lecture room	AGNSS	u-blox LEA-4P	27.8	42.5	30.8
Zhang	2011	lecture room	AGNSS	u-blox EVK-5H	21.9	32.5	20.6
Kjærgaard	2010	wood & brick	AGNSS	u-blox LEA-5H	4.0	6.0	40.0
Kjærgaard	2010	shopping mall	AGNSS	u-blox LEA-5H	15.0	18.0	> 1 min
Kjærgaard	2010	shopping mall	AGNSS	Nokia, GPS 5300	15.0	18.0	90.0
Eissfeller	2005	wood dome	massive parallel	SiRF-Star II	16.1	22.7	hot start
Eissfeller	2005	wood dome	massive parallel	SiRF-Star III	15.3	20.8	hot start
Soloviev	2010	up to -188dBW	carrier phase	L1 only GPS	cm	-	-

12 Pseudolites

The term 'pseudolite' is an accepted short form for pseudo-satellites, which are land-based beacons that generate pseudo-noise codes similar to those transmitted by GNSS. A pseudolite system also includes mobile receiver units (rovers) whose positions are estimated from distance measurements to the pseudolite beacons which are usually deployed at known positions. The main purpose of pseudolites is to support GNSS with additional ranges in situations where satellite signals are blocked, jammed or simply not available, e.g. in indoor environments. Originally, pseudolites included only systems that transmit at GPS frequencies L1 (1575.42 MHz) and/or L2 (1227.6 MHz) to enhance the satellite geometry for the usage of customary GPS receivers. Although utilization of the civilian signal structure of GPS is advantageous because existing receiver hardware can be used, legal broadcasting of GNSS signals is very limited. Therefore research and development has moved away from preserving backward compatibility with existing GNSS. For the upcoming Galileo system however, the GATE (2011) testbed consisting of 8 terrestrial transmitters fulfills that strict sense of a pseudolite system by broadcasting fully compliant Galileo signals – but it will operate only until Galileo has reached full operational capability in 2014. This chapter refers to a broader definition of pseudolites which includes all systems that have similarities to GNSS.

The coverage area of a pseudolite system can span up to tens of kilometers, with its limitation mainly driven by the availability of line of sight between pseudolites and rovers. Although local regulations set upper limits to the signal power, the received power from pseudolites is orders of magnitude stronger compared to the received signal strength of satellites. A combined pseudolite receiver can acquire and track GNSS and pseudolite signals. This functionality provides an extended positioning solution for seamless transition between indoor and outdoor environments.

Several error sources are intrinsic for pseudolite based indoor positioning:

- Multipath is a big concern when using signal structures similar to GNSS indoors. This problem contributes to the degradation of system performance. Multipath is mitigated by carrying out carrier phase measurements which are less sensitive to multipath, i.e. a typical multipath error of the carrier phase is in the order of

centimeters, compared to deviations in the order of 10 m for multipath code measurements (Kee et al. 2003).
- The near-far problem arises from signal interference of pseudolites with large differences in the distance to a receiver, causing such large disparities in received signal power that the weaker signal cannot be encoded. The problem is usually solved by pulsing the transmission of each pseudolite separately in time.
- Time synchronization remains a costly and complex task for pseudolite range measurements. This is particularly the case in deep indoor environments where access to atomic clocks of the GNSS satellites is not available.
- A solution for carrier phase ambiguities is not found straightforwardly for pseudolite systems relying on phase measurements. Methods for finding an ambiguity solution are triple differencing or keeping the receiver in motion while collecting carrier phase data in the initialization phase. Note that pseudolites – in contrast to satellites – don't move and therefore do not provide additional geometric constraints while the receiver is static.

12.1 Pseudolites Using Signals Different to GNSS

This group of pseudolite systems uses a dedicated signal infrastructure which is similar – but not identical – to that of GNSS. An advantage is that license-free frequency bands can be used with better prospects of commercialization. Such systems have an architecture that usually includes distance estimation by pseudo ranges and carrier phase measurements.

The Locata (2011) technology consists of a network of terrestrially-based and time-synchronized pseudolite transmitters which broadcast GNSS-like signals for kinematic applications at cm-level accuracy using carrier-phase measurements. Two frequencies within the 2.4 GHz ISM band are used for broadcasting. The Locata transmitters can be synchronized to an accuracy of 3 ns. Deployed Locata transmitters can augment GNSS positioning in situations where the reception of GNSS signals is severely degraded. Intended applications are deep pit mines, where high walls block a significant number of GNSS satellites. Operating ranges of up to 50 km have been reported. Rizos et al. (2010) have conducted indoor experiments in a hall with an area of 30 m by 15 m. After installing a network of 5 Locata transmitters, a standard deviation for static measurements of 2 cm was achieved. Barnes et al. (2005) demonstrate the suitability of the Locata technology for machine tracking and guidance in factories or warehouses where GNSS satellite coverage is limited. Barnes et al. (2007) conclude that movements of less than 1 cm can be detected. Since the signals can be transmitted with significantly larger amplitudes compared to GNSS, Locata signals can penetrate walls, but with a degraded level of performance.

Trimble (2011) offers a so-called Terralite XPS system consisting of ground-based transmit stations which are used in combination with GPS/GLONASS to extend coverage for a positioning service in open pit mines. Very little information about the system specification has been made public.

Kee et al. (2001 and 2003) describe a system of asynchronous pseudolites which have been shown to provide accuracy of 1 mm to 2 mm in static experiments, respectively 5 mm to 15 mm in dynamic indoor environments. These results were achieved using customary GPS receivers with minor software modifications that allowed observing differential carrier phase measurements to four pseudolites.

12.2 GNSS Repeaters

GNSS repeaters use – in contrast to pseudolites – the original, unmodified (but amplified) GNSS signal for positioning. The purpose of using repeaters is to enable customary GNSS receivers to track satellites in a blocked signal environment. The GNSS repeater approach includes a GNSS antenna located outdoors at a location with direct view to the satellites. A preferred location is the roof of a building. The received GNSS signal is transferred from the antenna into the building via cable and repeated wirelessly inside the building by multiple transmitters which consist of a signal amplifier and an internal rebroadcast antenna. If only a single transmitter is used for re-transmission, a mobile receiver in the coverage area will always determine its own location to be that of the static outdoor antenna. The extra path delay from the repeater cable is common to all satellites and thus indistinguishable from the receiver offset. To enable positioning indoors with higher resolution, multiple transmitters can be deployed. Based on a sequential broadcasting scheme the individual transmitters can be identified from fixed time slots within a transmission cycle. The advantage of such an approach is that the synchronization of the transmitters is automatically provided by GNSS time. At the rover, the position is determined by multilateration from delta distances (distance differences or pseudo distances) between the rover and the transmitters. These pseudo distances are obtained from an offset in the carrier phase of the received signal at the rover.

The implementation of a repeater system requires compliance with the regulations for its legal operation. The ECC Report-145 (2010) details the regulatory framework for GNSS repeaters. In order to prevent signal interference which can impact the normal use of GNSS in the vicinity, the regulations limit the total gain of +45 dB for GNSS repeaters, which is equivalent to a GNSS protection distance of around 10 m for a typical GNSS signal strength of -160 dBW. However, most countries currently do not allow using GNSS repeater equipment at all.

Fluerasu et al. (2011) present a proof-of-concept of a repeater system with multiple indoor transmitters. The stated accuracy of their implementation is 2 m to 3 m. In order to reduce the interference problem for uninvolved GNSS receivers located in the vicinity, the PRN codes 33 to 36 were used, which are reserved to support pseudolites.

12 Pseudolites

Vervisch-Picois and Samama (2009) propose a concept of so-called repealites (short for repeater-pseudolites), which represent a compromise between GNSS repeaters and pseudolites. The same satellite signal is retransmitted continuously onto all antennas of an indoor setting, but a distinct delay is applied to each antenna in order to avoid interference. From four delayed tracking channels, the receiver is able to determine its own position from four pseudoranges with an accuracy of 40 cm. If carrier phase measurements are used and the phase ambiguities can be solved, the authors predict that an accuracy of 10 cm to 15 cm is feasible.

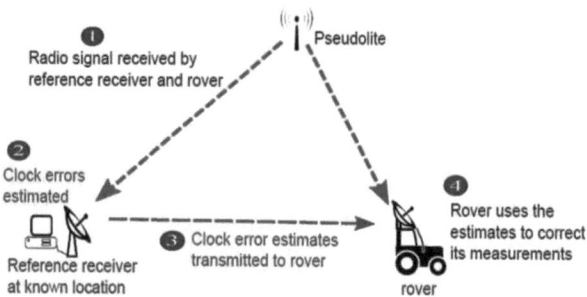

Figure 12.1 System overview of Alawieh et al. (2010)

The experimental system of Niwa et al. (2008) includes repeaters from a rooftop antenna and pseudolites transmitting GPS L1 codes with the PRN numbers 33 to 36 which are not used for satellites. Carrier phase measurements to locate a robot proved to be unreliable due to cycle slips which occur when the robot is near a wall.

Alawieh (2010) presents a method for modeling the clock drift of a pseudolite system which is currently under development at the Fraunhofer Institute IIS. The system architecture includes a special differential technique which employs additional reference receiver stations which estimate the clock bias of the pseudolites and forward the value as a time correction to a rover, see Figure 12.1.

12.3 Summary on Pseudolite Systems

Several difficulties have limited pseudolite systems to few applications in GNSS-challenged environments such as open pit mines. These difficulties arise primarily from the need for multipath mitigation, time synchronization and ambiguity solving. The tempting approach to broadcast the GPS L1 signal or to use GNSS repeaters is impeded by regulatory restrictions. Therefore, pseudolites of commercial systems always broadcast their own signal structure. While an area of several kilometers can be covered outdoors, the indoor application is limited to a single room or part of a building. An overview of pseudolite systems included in this study is given in Table 12.1.

12.3 Summary on Pseudolite Systems

Table 12.1 Localization systems using pseudolite technologies and reported performance parameters.

Name	Year	Principle	Reported Accuracy (m)	Coverage Area (m^2)	Frequency	Application	Market Maturity
Rizos (Locata)	2010	carrier phase ranges	2 cm	50 km	2.4 GHz	pit mines, also indoors	product
Trimble	2011	not disclosed	cm - dm	km	not disclosed	pit mines	product
Kee	2001	carrier phase ranges	1-2 cm	9 m^2	1.0 GHz	factories, indoor nav.	study
Fluerasu	2011	GNSS repeaters	2-3 m	building	1.6 GHz, L1	emergency, fire fighters	simulation
Vervisch-Picois	2009	repealities	< 1 m	400 m^2	1.6 GHz, L1	commercial indoor use	study
Niwa	2008	carrier phase ranges	cm - m	100 m^2	1.6 GHz, L1	robots for everyday life	study
Alawieh	2010	carrier phase ranges	4 cm	100×100 m^2	not disclosed	enhanced navigation	study

13 Other Radio Frequency Technologies

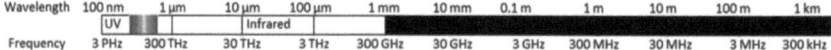

A major class of indoor positioning systems utilizes propagation of Radio-Frequency (RF) signals emitted by either dedicated beacon structures, notably Ultra Wide Band Radios (Chapter 10) and RFID tags (Chapter 9) or alternatively via common wireless communication standards via an existing network of deployed nodes such as WLAN (Chapter 8). This chapter discusses communication protocols such as ZigBee (Section 13.1) and Bluetooth (Section 13.2) which have – like WLAN – been designed for short-range wireless transfer. The existing infrastructure of short range DECT phones (Section 13.3) as well as wide-area digital television (Section 13.4) and Cellular Networks (Section 13.5) can also be used for indoor positioning. Systems based on RF signal reflection are discussed in Section 13.6 (Radar). Chapter 13 concludes with a discussion on positioning using long-range FM Radios (Section 13.7).

13.1 ZigBee

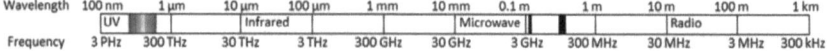

ZigBee is a wireless technology standard which can be regarded as a low rate Wireless Personal Area Network (WPAN). It is particularly designed for applications which demand low-power consumption, but don't require large data throughput. The signal range coverage of a ZigBee node is up to 100 m in free space, but in indoor environments typically 20 m to 30 m. Distance estimation between two ZigBee nodes is usually carried out from RSSI values. Since ZigBee operates in the unlicensed ISM bands, it is vulnerable to interference from a wide range of signal types using the same frequency which can disrupt radio communication.

Larrañaga et al. (2010) deployed a ZigBee network consisting of 8 reference nodes in an office space area of 432 m². RSSI values were used for locating a mobile ZigBee node. Instead of generating a map of fingerprints, the distances between the known positions of the reference nodes were used to obtain information about the real propagation characteristics of the scenario. An average localization accuracy of 3 m was achieved. In a similar study carried out by Tadakamadla (2006) it was concluded that the main error contribution is caused by the randomness of RSSI and the dependency on the user's orientation and the user's body. My-Bodyguard (2011) offers a tracking system for persons and objects based on ZigBee for indoor areas and GNSS otherwise. My-Bodyguard is supposed to deliver room-level accuracy at a measurement rate of 0.2 Hz. The beacons' battery power is specified for an active signal transmission duration of 3 years.

13.2 Bluetooth

Wavelength	100 nm	1 μm	10 μm	100 μm	1 mm	10 mm	0.1 m	1 m	10 m	100 m	1 km
	UV		Infrared			Microwave			Radio		
Frequency	3 PHz	300 THz	30 THz	3 THz	300 GHz	30 GHz	3 GHz	300 MHz	30 MHz	3 MHz	300 kHz

Bluetooth is – like ZigBee – a wireless standard for Wireless Personal Area Networks (WPANs). But in contrast to ZigBee, the Bluetooth standard is a proprietary format managed by the Bluetooth Special Interest Group. The advantage of using Bluetooth for exchanging information between devices is that this technology is of high security, low cost, low power and small size. The highest specified power level (class 3) of the Bluetooth standard has a maximum power output of 1 mW (0 dBm) which enables communication ranges of 5 m to 10 m depending on the propagation conditions such as LOS, material coverage and antenna configuration. Since the Bluetooth sensor does not stay in inquiry mode for 5 s during its 10 s cycle, the off-the-shelf Bluetooth device has latency unsuitable for real-time positioning applications.

According to Cheung et al. (2006) the consensus in published work on Bluetooth positioning implies that the standards and intrinsic characteristics of the protocol do not favor conventional time-of-flight based positioning methods. Reading RSSI for an unmodified standard Bluetooth device is not in option either, since the host controller command for reading the received signal strength values is not implemented by default (Aalto et al. 2004). Therefore, the Cell of Origin (CoO) method is normally applied as the basic positioning principle. Aalto et al. implemented the Real-Time Navigational Assistance (URNA) system to enable the transfer of location-based information between Bluetooth-enabled mobile phones. Using CoO, the position accuracy was stated as 10 m to 20 m, which was the achievable range of device discovery during tests carried out in a corridor. URNA is designed for location-aware mobile advertising to mobile phones.

Being aware that RSSI is not a reliable measure of Bluetooth hardware, Bargh and de Grote (2008) used a fingerprint-based localization method which relies only on the Response Rate (RR) of Bluetooth inquiries. It was shown, that the measured RR decreases with respect to the distance, e.g. for 2 m RR = 97% and for 10 m RR = 86 % at variances of 1 % to 5%. An extensive fingerprinting capture was necessary to achieve sub-room accuracy.

13 Other Radio Frequency Technologies

ZONITH (2011) offers an indoor positioning module which consists of deployed Bluetooth beacons (each covering one or more rooms) and Bluetooth devices, such as those integrated in mobile phones, worn by the people to be tracked. On a graphical interface, it is possible to view the employees' locations. ZONITH provides room-level accuracy sufficient for applications of worker protection and automatic alarm dispatch.

13.3 DECT Phones

Wavelength	100 nm	1 μm	10 μm	100 μm	1 mm	10 mm	0.1 m	1 m	10 m	100 m	1 km
	UV		Infrared			Microwave			Radio		
Frequency	3 PHz	300 THz	30 THz	3 THz	300 GHz	30 GHz	3 GHz	300 MHz	30 MHz	3 MHz	300 kHz

Phones based on Digital Enhanced Cordless Technology (DECT) are common devices for talking wirelessly around the house. DECT phones communicate with a single base station within a typical distance of 200 m to 500 m.

Kranz (2010) demonstrated the feasibility of using DECT phones for positioning in urban indoor and outdoor scenarios. His fingerprinting method on DECT RSSI outperformed that of WLAN due to the high number of DECT stations (12 to 17) which could be received at a single sub-urban location. After taking fingerprints of 1 to 3 m separation, a localization accuracy of up to 5 m was achieved.

Schwaighofer et al. (2003) have taken a dense grid of 650 RSSI fingerprints from typically 15 surrounding DECT phones in an assembly hall of 250 m × 180 m. Due to dynamical changes caused by moving cranes and people, the RSSI fluctuated by 10 dB. An average deviation of 7.5 m was achieved by building Gaussian process models for the RSSI distribution.

13.4 Digital Television

Wavelength	100 nm	1 μm	10 μm	100 μm	1 mm	10 mm	0.1 m	1 m	10 m	100 m	1 km
	UV		Infrared			Microwave			Radio		
Frequency	3 PHz	300 THz	30 THz	3 THz	300 GHz	30 GHz	3 GHz	300 MHz	30 MHz	3 MHz	300 kHz

Broadcast signals of digital television stations can be utilized for positioning in urban areas including deep indoor environments with accuracy of about 10 m. Since digital television had started in 1998, most countries have established a network of terrestrial broadcast stations (10 km to 100 km distance between stations). The unmodified digital video broadcast is suited for pseudorange estimation and multilateration in indoor environments due to several reasons:

- digital television has a signal power advantage over GPS of 40 dB allowing reception in deep indoor environments
- the signals have a wide bandwidth of 5 MHz – 8 MHz facilitating multipath mitigation
- demodulation of the data is simplified by a guard interval in the message
- emitters of digital television are synchronized with GPS time allowing to timestamp the data and determine TOA pseudoranges.

However, the weak density of terrestrial emitters is low – causing the direct signal to arrive at low elevation angles near the horizon. With such a network configuration, only 2D positioning is feasible and multipath is severe since the direct signal is usually blocked.

Rabinowitz and Spilker (2005) have shown that the infrastructure of digital television can be exploited for obtaining positions in parking garages and in office buildings. In the ground floor from an office building an accuracy of 10 m was achieved.

Serant et al. (2011) describe a pseudorange estimation method based on digital video broadcast signals and demonstrate that pseudoranges can be determined with an accuracy of 20 m even in the presence of NLOS signals and blockage of the direct signals. With advanced processing techniques the research team was able to improve these results to below 10 m.

13.5 Cellular Networks

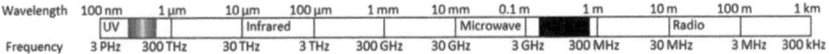

This section is dedicated to positioning techniques which solely rely on the mobile cellular network, notably the second-generation wireless telephone technology GSM (Global System for Mobile communication). It does not include navigation techniques which make use of sensor components being integral part of smart phones such as INS (MEMS accelerometers, gyroscopes, and magnetometers), GNSS, cameras and WLAN. See Chapter 14 for solutions based pm INS-supported pedestrian navigation.

Mobile networks such as GSM and Universal Mobile Telecommunications System (UMTS) form the basis of modern area-wide wireless communication infrastructure. GSM networks are pervasively available in most countries. With cell sizes of up to 35 km, GSM far outreaches the coverage of WLAN. Therefore, obtaining position estimates for the mobile user – even with low accuracy – sees mass market applications such as location based services and emergency assistance. For locating a mobile phone, it is not necessary to make an active call. Unlike WLAN, GSM operates in the licensed bands, where there is no interference from other devices operating at the same frequency.

All mobile phone tracking methods have in common that they use the locations of nearby antennas mounted at radio towers to infer the mobile's position. These are: RSSI fingerprinting, RSSI distances, AoA, ToA, TDoA and CoO (see Section 3.3 for an explanation). Ingensand and Bitzi (2001) describe the common methods for GSM positioning and show the advantages and disadvantages of these methods. The achievable accuracies are listed and considerations on the improvement of accuracy are made. A detailed theoretical background of the common methods for mobile positioning including the derivation of fundamental performance bounds can be found in Gustafsson and Gunnarsson (2005).

13.5.1 GSM Fingerprinting

The most common method of GSM indoor locating is fingerprinting based on the power level (RSSI). GSM fingerprinting has two advantages: First, the poser level observed by a mobile device exhibits a significant variability at around 1 m to 10 m in indoor spaces. Secondly, the hardware of customary mobile phones can be used. The shortcomings of GSM fingerprinting are low reliability due to varying signal propagation conditions and the need for an area-wide, but also dense grid of fingerprints. Even for indoor positioning the GSM signal travels mostly outdoors and is affected by changing weather, foliation and construction (Popleteev 2011). In contrast, Varshavsky et al. (2007) show that the GSM signal is extremely stable over time. Note that RSSI based localization is not possible for CDMA networks where the transmission power is dynamically adjusted at the radio towers according to the current needs.

Varshavsky et al. (2007) conducted experiments on RSSI fingerprints collected in multi-floor buildings. By reading up to 29 GSM channels at a time and taking GSM fingerprints 1 m to 2 m apart, they were effectively able to differentiate between floors of the building. Depending on the number of measurements per location as well as the number of received channels and data collection grid size, they achieved a positioning accuracy of 2 m to 4 m.

13.5.2 Distance Based GSM Positioning

The project report of SoLoc (2009) describes the difficulties that arise from distance-based positioning on GSM. First, phone providers keep their locations concealed. Second, it is difficult to find a suitable signal propagation model adequate for various scenarios. SoLoc used the Walfisch-Ikegami model which considers building heights, antenna heights and even street widths. Gustafsson and Gunnarsson (2005) demonstrate that accuracy of 100 m is realistic for distance-model based GSM positioning.

Using two locally deployed sources of GSM cellular signals, Loctronix (2011) claims to achieve meter-level accuracy based on Doppler ranging. Apart from dedicated GSM beacons, the system can make use of various signals of opportunity such as those from GSM cell towers, digital television, WLAN and GNSS. Depending on the availability of signals, the system achieves accuracies of 1 m to 15 m. At the time of writing, products were not yet on the market.

13.5.3 Angle Based GSM Positioning

The resolution in AoA (Angle of Arrival) depends on the antenna configuration. GSM base stations are typically equipped with directionally sensitive antennas which provide crude angle information, e.g. 120° angles for a 3-sector antenna. Some radio towers equipped with group antennas establish sectors of 30°. For the indoor case, AoA information is further diluted due to the presence of NLoS and multipath propagation.

The Enhanced Cell of Origin (ECoO) method has been specially designed for mobile phones, which refers to a cell-sector positioning method to provide angular information and distance estimation to cell towers by measuring the Round Trip Time (RTT). In case of Line of Sight (LoS), relatively good distance accuracy can be achieved by RTT, but LoS conditions are not common in indoor environments. Shen and Oda (2010) showed in their simulations that ECoO could be used to estimate direction with an accuracy of 11° in a Rayleigh fading environment, i.e. assuming NLoS conditions. Using ECoO and RTT measurements, they predict a 2D position accuracy of 100 m to 150 m at a cell tower spacing of 4 km.

13.6 Radar

Wavelength	100 nm	1 μm	10 μm	100 μm	1 mm	10 mm	0.1 m	1 m	10 m	100 m	1 km
	UV		Infrared			Microwave					
Frequency	3 PHz	300 THz	30 THz	3 THz	300 GHz	30 GHz	3 GHz	300 MHz	30 MHz	3 MHz	300 kHz

Radar (RAdio Detection And Ranging) is a technique to determine the range and angle of incidence to an object. The original principle of radar was to measure the propagation time and direction of radio pulses transmitted by an antenna and then bounced back from a distant passive target (primary radar). If the object returns a tiny part of the wave's energy to the antenna, the radar device measures the elapsed time. The angle of incidence is estimated from a directional antenna. This original concept of radar assumes passive object reflection and involves only one station which comprises both, transmitter and sensor. This concept has two disadvantages: most of the signal energy gets lost by the reflection and the use of steerable directional antennas is impracticable. Therefore, the concept of radar has been extended to include more than one active transmitter (secondary radar). Instead of passive reflection, the single-way travel time of the radar pulse is measured by ToA and then returned actively. Some methods such as FMCW Radar (see 13.6.1) transmit a signal continuously. If the transmissions in each direction are sufficiently separated in frequency, the return can occur immediately. Instead of using directional antennas, the angle of incidence can be determined indirectly by multilateration from several transmitters at fixed positions. The two concepts have in common that the Round Trip Time (RTT) is linearly related to the distance.

13.6.1 FMCW Radar

Frequency Modulated Continuous Wave (FMCW) radar is a short-range measuring technique, where the transmitter frequency is linearly increased with the time. The returned echo is received with a constant offset Δt, which relates to the travelled distance, see Figure 13.1. An advantage of FMCW is its resistance to the Doppler effect. The Doppler movement only introduces a shift in the frequency which is canceled out by differencing.

Appling the FMCW technique for distance measurements between two devices, accuracies in the order of a few centimeters can be achieved. Most FMCW Radar implementations make use of multilateration based on RTT distance estimates between a mobile transmitter and multiple fixed transponders. The transponder broadcasts a radio signal in the free ISM-band (5.725 GHz to 5.875 GHz) which is received, processed and echoed back to the transponder by each transmitter without time delay. The echo is coded with the respective transponder's identification in order to allow the transmitter to separate each transponder's answer.

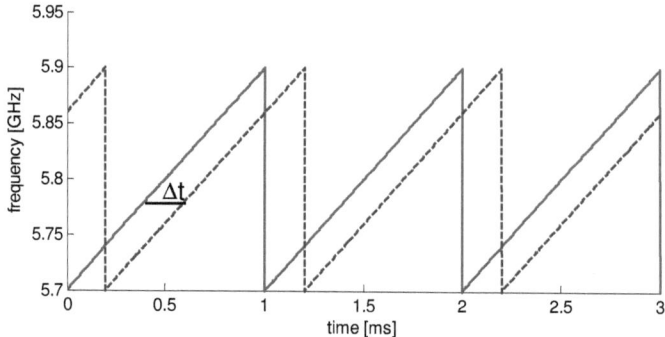

Figure 13.1 Linear frequency increase of transmitted signal (solid). Received echo (dashed). The constant time delay Δt relates to the distance.

Stelzer et al. (2004) have prototyped a local positioning system based on FMCW radar with a system architecture of multiple fixed base stations and a lightweight mobile transponder operating at 5.8 GHz. Based on TDoA ranges at cm-level precision measured under LoS conditions, a positioning accuracy of 10 cm over 500 m has been achieved. At a measurement rate of 1000 Hz, the system can be used to track fast vehicles at race courses.

Symeo (2011) offers the 2D local positioning system 'LPR-2D' which can be used to determine the position of a mobile transponder within accuracy of up to 5 cm. At least 3 distance measurements to 3 fixed reference transmitters mounted in a distance of up to 400 m are required. Combined indoor/outdoor use is possible from an integrated differential GNSS receiver. The system is dedicated for operation under harsh industrial conditions to guide vehicles and cranes. A detailed system description has been published by Röhr and Gulden (2009). ABATEC (2011) offers the 3D positioning system 'LPM' based on FMCW radar for various applications such as sport analysis and infotainment.

13.6.2 Chirp Spread Spectrum (CSS)

CSS is a spread spectrum technique similar to FMCW (13.6.1) which uses wideband linear frequency modulated chirp pulses instead of continuous waves. A chirp of CSS is a sinusoidal signal whose frequency increases and decreases over a certain amount of time. Figure 13.2 shows an example chirp. In contrast to FMCW the chirp pules have a fixed duration of 1 μs followed by an interval with no signal, which reduces the required transmission power.

The main advantage of using CSS is robustness to noise and multipath owing to the wide bandwidth used to produce a chirp signal. Precise time synchronization between the devices is not required. The chirp pulses can be used for range measurements with low power consumption. In contrast to FM radios (Section 13.7) which operate between 88 MHz to 108 MHz, CSS is used in the 2.45 GHz band. Nanotron Technologies (2011) has marketed an RF positioning kit which uses a network of up to 16 nodes that can either be configured as beacon or mobile node. Positioning is carried out on CSS ranging at an accuracy of 1 m. Solcon (2011) offers a similar location system based on CSS frequency modulation. The distances are determined by Symmetrical Double-Sided Two Way Ranging (SDS-TWR), a ranging methodology that uses two delays.

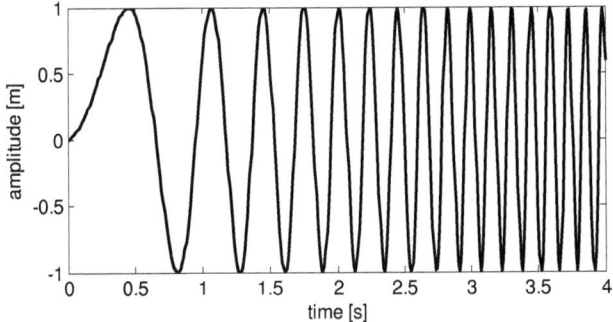

Figure 13.2 Frequency modulated chirp, $a(t) = \sin(2\pi t (0.1 + t))$

13.6.3 Doppler Radar

Doppler radar is a special radar technique for the detection and tracking moving targets. It is common for airborne applications but can also be applied for infrastructure free indoor navigation. Doppler radar utilizes Doppler ranging, i.e. relative velocity measurements from the Doppler shift between an emitted radar wave and the return signal after reflection from an obstacle (e.g. a wall). It can only be used to determine the radial component of a target's velocity.

Yokoo et al. (2009) use two continuous wave Doppler radars to track a vehicle. Without employment of other sensors, the positioning technique of Doppler radar shows a strong drift due to unstable orientation (2 m deviation for a 20 m trajectory). Nevertheless, when supporting the Doppler radar with data from a gyro sensor the accuracy improved by a factor of 30 (6.8 cm instead of 2 m). The intended application is pedestrian navigation for first responders.

13.6.4 Wave Field Analysis

The principle of distance measurement is to establish a 'standing wave' between the transmitter and receiver unit in order to determine the phase shift. Several such phase measurements are carried out sequentially for a discrete set of frequencies in the spectrum between 2.4 GHz to 2.48 GHz. A change of 1 MHz for example, causes a phase shift of ca. 24° at a certain distance, but will be 48° for the doubled distance. The final distance is than determined by Fourier analysis. The advantage of such a phase measurement technique is that there is no need for accurate synchronization or time measurements.

The company 'lambda: 4 Entwicklungen GmbH' (Lambda:4 2011) is prototyping a positioning system using wave field analyses for the purpose of rescuing people buried by an avalanche or locating fireman in a hazardous environment. The system measures distances between a handheld locating device of approximately 1.2 kg and an active transmitter of matchbox size. Positions are determined from multiple devices via multilateration or direction estimation using RSSI from an antenna array. The system is capable of detecting the tags with an accuracy of less than 1 m for the LoS case and 1 m to 5 m accuracy in the case of NLoS, i.e. walls in the direct signal path. Under LoS conditions it can cover distances up to 100 m. According to Reimann (2011), a static system with constant antenna orientation can achieve millimeter accuracy.

13.7 FM Radio

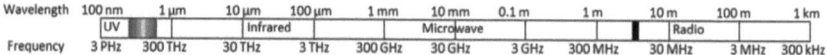

FM radios represent a well-established broadcasting technology originally reserved for Frequency Modulation (FM) to convey information over a carrier wave by varying its frequency. FM nowadays refers to any radio wave operating in the frequency band 87.5 MHz to 108.0 MHz no matter what type of signal modulation is applied. The audio signals transmitted by FM radio broadcasting towers can be used for indoor navigation. ToA and TDoA methods have not been regarded as feasible because the FM signal lacks timing information. But signal strength based fingerprinting techniques which are independent of clock synchronization can be applied on broadcast FM signals. The employment of FM radios for positioning benefits from the existence of radio tower infrastructure providing almost ubiquitous coverage in indoor and outdoor environments, high received signal power and the possibility of using low-cost, low-power hardware. Due to the passive nature of the client devices, FM can be used in sensitive areas where other RF technologies are prohibited for safety or security reasons. On top of that, FM technology is readily available in many mobile devices.

Moghtadaiee et al. (2011) have implemented an RSSI fingerprinting positioning system based on FM radios in an office environment. After obtaining FM fingerprints consisting of 17 sensed FM channels at 150 grid points within an area of 253 m², a mean accuracy of 3 m could be achieved. However, under the same conditions, the WLAN fingerprinting showed higher accuracy.

In contrast to Moghtadaiee et al., Popleteev (2011) concluded in his PhD thesis that indoor positioning based on broadcast FM stations is comparable to WLAN in terms of accuracy and outperforms other RF technologies in battery life and coverage. Popleteev carried out RSSI fingerprinting on a 1 m grid of calibration points. Using RSSI values from 76 active FM broadcasting stations and 3 local, self-installed FM transmitters, he achieved a positioning accuracy of 5 m (1 m median). Papliatseyeu et al. (2009) rely exclusively on short-range, self-deployed FM transmitters. Their fingerprinting method achieved similar results of 5 m accuracy.

13.8 Summary on Radio Systems

Any radio signal can be used for indoor positioning at any frequency, signal range and protocol, see Table 13.1. However, performance levels and applicability vary greatly depending on several factors such as the use of pre-existing reference infrastructure, pervasiveness of devices, signal ranges, power levels etc. Signal strength fingerprinting and distance based methods can be considered. While fingerprinting at meter level accuracy is impracticable due to the required pre-calibration, time-of-flight measurements remain unreliable indoors due to complicated signal propagation. The outstanding sub-millimeter accuracy which interferometry offers is only of relative type and therefore hardly applicable for indoor navigation.

Table 13.1 Localization systems using various RF technologies and reported performance parameters

Name	Year	Wireless Technology	Reported Accuracy (m)	Area (m^2) or No. of Reference Nodes	Principle	Prior calibration	Application	Market Maturity
Larrañaga	2010	ZigBee	3 m	432 / 8 = 54	RSSI distance	minimal	context, LBS	study
Tadakamadla	2006	ZigBee	3 m	192 / 5 = 48	RSSI distance	yes	WSN, tracking	study
My-Bodyguard	2011	Zigbee	room	1 per room	CoO	no	tracking	product
Aalto	2004	Bluetooth	20 m	2E4/9= 2222	CoO	no	advertising	study
Bargh	2008	Bluetooth	room	sub room	fingerprinting	yes	LBS	study
ZONITH	2011	Bluetooth	room	1 per room	CoO	no	employee tracking	product
Kranz	2010	DECT phone	5 m	900 / 12 = 75	fingerprinting	yes	seamless positioning	study
Schwaighofer	2003	DECT Phone	7.5 m	4E4 /15 =3E3	fingerprinting	yes	LBS	study
Rabinowitz	2005	Digital TV	10-20 m	15 km^2	pseudoranges	no	emergency response	study
Serant	2011	Digital TV	10-25 m	80 km^2	pseudoranges	no	urban LBS	study
Varshavsky	2007	GSM	4 m	building / 29	fingerprinting	yes	indoor localization	study
Loctronix	2011	GSM	1-15 m	325 / 2= 176	Doppler ranges	no	LBS on cell phones	product
Shen	2010	GSM	150 m	12km^2/tower	ECoO, RTT	no	mobile phone LBS	simulation
Stelzer	2004	FMCW radar	10 cm	500 x 500 m^2	multilateration	no	race track	study
Symeo	2011	FMCW radar	5 cm	4E4/3=1.3E3	RTT, beacons	no	forklift tracking	product
ABATEC	2011	FMCW radar	3 cm	1 km^2	multilateration	no	sport analysis	product
Nanotron	2011	CSS	1 m	2-16 beacons	ToA, beacons	no	loss protection	product
Solcon	2011	CSS	meter	> 2 per room	TWR, beacons	no	facility management	product
Yokoo	2009	radar	0.3 %	25 m^2	Doppler radar	no	first-responders	study
Lambda:4	2011	radar	1 m	1 or more	lateration, RSSI	no	avalanche rescue	product
Moghtadaiee	2011	FM Radio	3 m	radio stations	fingerprinting	yes	indoor navigation	study
Popleteev	2011	FM Radio	5 m	radio stations	fingerprinting	yes	employee tracking	study
Papliatseyeu	2009	FM Radio	4.5 m	72 / 3 = 24	fingerprinting	yes	employee tracking	study

14 Inertial Navigation Systems

This chapter outlines Pedestrian Dead Reckoning (PDR) approaches based on Inertial Navigation Systems (INS) consisting of an Inertial Measurement Unit (IMU) and a processing unit as the main components. Since an INS has a significant drift, it is usually fused with complementary sensors which provide absolute location information.

An INS is an electronic device which provides estimates of position, velocity and orientation from an IMU. The custom IMU consists of three orthogonally arranged accelerometers (motion sensors), three gyroscopes (angular rate sensors) and/or a magnetometer (3 perpendicular sensors for measuring the strength and/or direction of a magnetic field). If the initial position and orientation are known, subsequent positions, orientations and velocities (direction and speed of movement) of the moving platform can be updated continuously via Dead Reckoning (DR) without the need for external reference positions. The main argument to use INS for pedestrian navigation arises from independent operability – at least temporarily – without external infrastructure, making navigation possible in environments, where the installation and maintenance of such infrastructure is not affordable. Comprehensive information about the modes of operation and applications of an INS is given by Jekeli (2001).

14.1 INS Navigation without External Infrastructure

Using DR, an INS-based guidance system is continually adding detected changes to its previously-calculated positions. Since DR is only sensitive to changes, the INS can only measure relative position and orientation. Thus, the accuracy of the propagated position depends heavily on the quality of the provided start position and direction. Most notably, owing to the double integration of noisy accelerometer measurements, INS suffer from accumulation of position deviation and magnification of the angular deviation over the travelled distance (Abbe error). The shorter the integration time the higher is the predicted position from an INS. If absolute position or orientation updates are obtained by another sensor source at a high rate, the INS can be used to deliver positions with much higher precision compared to geometric interpolation between supporting points.

In pedestrian navigation, the accumulating positioning deviation of DR origins from two major types of deviations: the along-track deviation mainly caused by the step length imprecision and the cross-track deviation mainly caused by the imprecision of the azimuth measurement. While the accuracy deteriorates over a long operation time by variance accumulation, the INS can provide – as mentioned above – very high short term accuracy and detailed shapes of the travelled route. Therefore dead reckoning based on INS is optimally utilized for providing short-term solutions in shadow or transition areas which are not covered by an absolute positioning system.

Nowadays the utilization of Micro Electro-Mechanical System (MEMS) sensors allows integrating all system components in a mobile smart phone (Lukianto et al. 2010). The miniaturization is keeping costs, power, weight and required space to a minimum. Therefore MEMS based INS sensors are particularly efficient for the purpose of pedestrian navigation. Note that 3500 € are considered 'low-cost' compared to high precision INS at a cost of up to 150'000 €. Since an INS operates self-contained the availability and continuity of service are usually guaranteed. Most critical for the applicability of an INS are the performance metrics accuracy and integrity. Accuracy of pedestrian navigation systems is usually indicated in the ratio between deviation and covered distance or elapsed time. The ratio expressed in percent deviation of travelled distance varies greatly between 0.1 % and 20 % depending on the used methodology and employment of complementary sensors. In Table 14.1 various INS-based navigation approaches and their components are characterized.

Mathiassen et al. (2010) built a low-cost test IMU supported with observations from magnetometers and a barometer. Magnetometers and barometers are not of high accuracy and reliability, but provide absolute quantities which can support a self-contained IMU. While simulated variance propagation predicts 10 m drift deviation within 60 s, first real-world tests yielded 65 m deviation within the same time interval based on a stationary unit. Before such a low-cost system becomes applicable, the gyroscope drift – as the main error source – needs to be improved.

14.2 Pedestrian Dead Reckoning

From continuous updates of the walked distance and heading, the position can be propagated by Dead Reckoning (DR). The fundamental mechanism of pedestrian dead reckoning reads

$$\begin{bmatrix} x_k \\ y_k \end{bmatrix} = \begin{bmatrix} x_{k-1} + l_k \sin(\theta_k) \\ y_{k-1} + l_k \cos(\theta_k) \end{bmatrix}, \tag{14.1}$$

where l is the step length, θ the current heading estimate and (x, y) the coordinates in the horizontal plane. Index k is an abbreviated form of $t(k)$ denoting the number of a discrete point in time.

14.2.1 Step Length Estimation

A pedestrian's accelerometer output can be used for step detection, step count and step length estimation. For this purpose it is necessary to first categorize the human physical activity into walking, running, ascending/descending stairs, taking an elevator, standing still and irregular movements. Different dynamics in the movements can be recognized from accelerometers integrated in a mobile device which is carried on the waist or held in the hand. Where and how the device (i.e. the IMU) is carried by the person is decisive for the processing and affects the performance. Foot mounted devices allow applying additional constraints (such as zero velocities while walking) and special algorithms. Therefore, foot mounted approaches are discussed separately in Section 14.4.

Sun et al. (2009) make use of the waveforms of all three acceleration components to classify relevant pedestrian activities. From the acceleration signal, deduced quantities such as velocity, distance, signal magnitude, mean and variance can be derived and analyzed for activity labeling. Bao and Intille (2004) identified different activities with a success rate of ca. 90 %. Shin et al. (2009) demonstrated the importance of behavior classification (such as sit-down, walk and run) for IMU-based pedestrian navigation. Context awareness of a person can be obtained by a step length estimator improving the navigation performance of a system significantly. The steps are detected by identification of periodicities, typically by filtering the signal magnitude of the pedestrian's acceleration. In order to determine the travelled distance the steps are counted and the step lengths are estimated from the step frequency. As shown in Figure 14.1, the step length increases linearly with the step frequency. In addition, the acceleration variance can also give an indication of the step length, see Figure 14.2. When an absolute position is available, the step length can be recalibrated during the walk. Such an online calibration technique adapts very well to a certain context or a new situation.

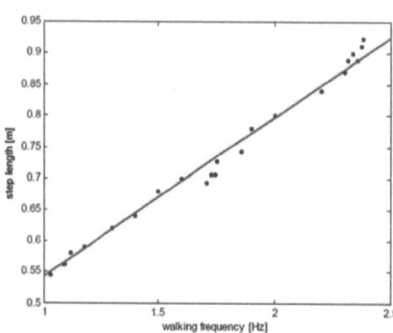

Figure 14.1 Step length in relation to the step frequency according to Shin et al. (2009)

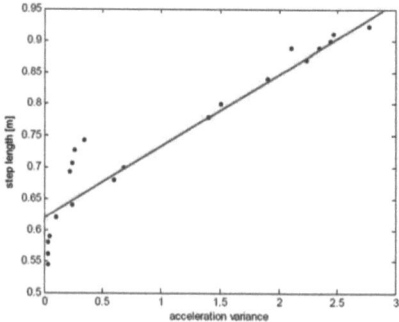

Figure 14.2 Acceleration variance versus step length, Shin et al. (2009)

14.2.2 Step Heading Estimation

The heading can be determined with azimuth readings of a leveled compass with a typical accuracy of 5° (Gusenbauer et al. 2010). The reliability of compass readings is compromised by magnetic perturbations caused by power lines or iron reinforcement and misalignment of the device (i.e. the line of sight of a pedestrian does not correspond to the direction of walk). A more reliable and therefore more common approach for direction estimation is the use of a gyroscope.

14.3 INS Pedestrian Navigation Using Complementary Sensors

Since the position deviation of dead reckoning grows with the time and distance traveled, position updates from an external source are necessary from time to time. Since in indoor environments GNSS are often severely degraded, the required external positioning information is typically obtained by beacon based systems such as WLAN, UWB or ultrasound. In principle any other absolute positioning system can serve this purpose. The pedestrian navigation system can help to navigate through areas where the complementary system has poor or no coverage.

Integration of the INS with a complementary system can have the form of loose coupling up to ultra-tight coupling. In a loosely coupled system the external positioning source is autonomously calculating a navigation solution and afterwards integrated in the INS solution. Ultra-tight coupling combines the raw observations into a navigation filter and makes use of a tracking feedback loop.

The main challenge in the implementation of hybrid localization algorithms is the formulation of the process and the measurement models. By applying Kalman or Particle filters, data from a large number of sensors can be fused such that the drawbacks of each sensor can be compensated with information from other sensors.

Sensors for pedestrian navigation can be subdivided into the two main classes: a) infrastructure independent sensors and b) sensors which require deployed local infrastructure. The second important criterion for subdivision is whether the sensor can measure an a) absolute quantity or b) relative changes to the target quantity. E.g. an accelerometer measures the absolute acceleration, but since position or orientation angles are the target quantities, accelerometers belong to the class of relatively measuring sensors in this context.

If an IMU does not take advantage of local infrastructure, the error growth resembles a quadratic function with the elapsed time (Jekeli 2001). In order to increase long-term stability, the position estimates need to be supported by a complementary system providing absolute position and possibly orientation references. Sensors employed to deliver absolute position information are typically GNSS (for 3D) and barometric pressure sensors for the height component. Apart from GNSS, any other sensor technology described in this report can be used (see Sections 14.3.2 to 14.3.5). The absolute orientation reference can be obtained from compasses (azimuthal orientation), magnetic field sensors (all three dimensions) or inclination sensors (two-axis horizontal orientation). Alternatively, the orientation can be derived from the position track if absolute positions (such as those obtained from GNSS) are available.

14.3.1 Combination of INS and GNSS

Most pedestrian navigation approaches include GNSS receivers in their system architecture. The advantage of a combined technique is that pedestrian dead reckoning and GNSS do not require local deployment of infrastructure. Therefore, the fusion of inertial navigation and GNSS allows for infrastructure-free positioning, which is a requirement for many applications, such as for rescue, fireman and soldiers. Since GNSS provides an absolute initial position and direction when entering a building, it is a crucial component for any INS-based indoor application. GNSS is also used for calibration and validation of pedestrian navigation systems, e.g. Beauregard (2006) who mounted the sensors on a helmet or Sun et al. (2009) who preferred attaching the system on a pedestrian's belt. Renaudin et al. (2007) mounted an AGNSS receiver on a pedestrian's cap and fused the GNSS observations with three inertial systems distributed over the body. However, the deterioration of AGNSS indoors (Chapter 11.2) causes the system to rely solely on INS while the pedestrian walks in a building.

14.3.2 Combination of INS and AoA

Kemppi et al. (2010) accommodate absolute location estimates from an angle-based localization system. The azimuth and elevation angles of the mobile receiver are determined by direction-of-departure principle from a multi-antenna array. A building map is also used for map matching and fused the with the IMU and angle measurements. Kemppi et al. report a deviation of 7 m after walking 2.5 minutes (190 m) and a deviation of 15 m subsequent to a walk of 6 minutes (270 m) using a single antenna.

14.3.3 Combination of INS and Optical Measurements

Aufderheide and Krybus (2010) propose a dual-track system combining inertial and optical measurements as a loosely coupled system. The camera pose is estimated from corresponding image features between successive frames from a monocular video camera. Therefore, their approach is also categorized as a camera system with reference from image sequences (Chapter 4.2). The knowledge of the camera position is used to bound the drift of the inertial system for long track durations. Conversely, pose estimations from IMU data limit the search space for feature tracking. Performance assessment is not yet possible because the system is currently under development.

Keßler et al. (2011) use a rotating laser scanner and alternatively a monocular camera to aid their pedestrian navigation system. The scan data is used to generate a 2D map and navigate within this map. Geometrical constraints inside buildings (e.g. orthogonality of structures) are used to recognize if a place is revisited. From the camera images the central vanishing points are determined to reduce the heading drift of the IMU. Complementary to the vision-based attitude aiding method, the position can be directly determined from comparison of images in a database. Tests with closed loops, which have been walked twice, show consistent paths.

The SLAM (Simultaneous Localization And Mapping) approach of Liu T. et al. (2010) combines IMU, laser scanner and image based localization, all integrated in a human-operated backpack system which can be used to generate 3D models of complex indoor environments. The positions are determined from data capture based on two laser scanners and an IMU with 6 Degree of Freedom (DoF). An average positional error of 1 % of the travelled distance is reported. The localization performance can be improved by making use of the camera images which have been taken in an offline phase. The images can be used to refine the six parameters of the camera pose and improve the quality of the 3D textured model.

14.3.4 Combination of INS and RSSI

Seitz et al. (2010) fuse low-cost accelerometer data with WLAN RSSI fingerprinting measurements using the Hidden Markov Model. Their pedestrian navigation system is dedicated for low-cost, low-power sensors integrated on smart mobile phones and achieves 5 m accuracy in indoor environments. Fink et al. (2010) also fuse the INS motion vector with radio localization using received signal strength. Their pedestrian navigation system is designed for the detection of maintenance staff in longwall mining, which is an underground coal mining technique. With the combination of WLAN path-loss distances and INS readings they report a position accuracy of 0.7 m – in contrast to a solution on WLAN fingerprinting of 5 m accuracy. With such an improvement the security zone can be reduced and the mining can be carried out more efficiently.

Apart from WLAN, it is also possible to use a network of ZigBee nodes at known locations to perform fingerprinting in combination with a pedestrian navigation system. Schmid et al. (2010) predict that such a system has the potential to provide 3 m accuracy within buildings.

14.3.5 Ultra Hybrid Systems

Klingbeil et al. (2010) present a modular fusion system for pedestrian navigation or robotic applications. The sensor platform includes common self-contained sensors worn on the person. For reference position updates GNSS can be used but it is also possible to utilize any other technique as reference such as deployed ultrasound nodes or radio frequency beacons. The system architecture is set up in such a way, that in principle all sensor modalities can be employed as 'control input' or brought in as observations for a particle filter. The processing is based on a Bayesian recursive estimation algorithm.

14.3.6 Combination of INS and Map Data

The required absolute position updates do not necessarily need to origin from external deployed infrastructure. Alternatively, detailed map data can be used, known as map measurements. If available indoor maps such as microMap or a CityGML are stored locally in the device, self-contained navigation becomes feasible. For correction of position and orientation, constraints from a map can be compared with the movement pattern to create complementary geometric information. This process is known under the term Map Matching (MM) and used to correct the current position track onto a map. The map matching approach exhibits a powerful backup strategy for correcting the position of the Dead Reckoning (DR) system. While car-roadmap matching is already an established tool in outdoor navigation, the less constrained movement pattern of a person requires special map matching algorithms tailored to pedestrian navigation.

Vertical displacements of pedestrians in buildings have been used by Gusenbauer et al. (2010) to identify an elevator or ascending/descending stairs. Wagner et al. (2010) use a topological Map Matching (MM) algorithm to feed back a Kalman Filter when GNSS positions are not available. Walder and Bernoulli (2010) propose the use of a building information model for their map matching approach. The model provides two important constraints to a pedestrian's track: a) the polygons of the outline of accessible areas and b) transition objects such as doors or stairs.

Pressl and Wieser (2006) present a pedestrian navigation system tailored to the needs of visually impaired people in an urban environment. By integrating IMU, GNSS and map matching the system achieves an absolute accuracy of 1 m to 2 m.

14.4 Foot Mounted Pedestrian Navigation

The installation of an IMU in a shoe has the purpose to mitigate the inertial drift. Foot mounted pedestrian navigation systems benefit considerably from supporting the IMU integration process with a Zero velocity-UPdaTe (ZUPT) or a Zero-Angular Rate Update (ZARU). During the foot is in stance stage the velocity estimation is recalibrated from the integrated acceleration with a zero value. These updates can be used as observation input to the Kalman filter in order to efficiently reduce the error accumulation of an IMU. The difficulty of the ZUPT approach is the detection of the moment of zero velocity under changing conditions such as various types of motion and speeds or different floor cover.

The ZUPT method allows replacing cubic in time error growth (t^3-drift) with an error accumulation linear to the number of steps (Wan and Foxlin 2010). Figure 14.3 illustrates how velocity updates circumvent a cubic error growth. The zero velocity updates can be introduced into the EKF (Extended Kalman Filter) as velocity measurements, with the purpose to deduce and correct the drift deviation (Figure 14.3b). The comparison of current pedestrian navigation systems in Table 14.1 shows that a drift of less than 1 % can be achieved by foot mounted systems, compared to IMUs attached to other body parts with drifts typically exceeding 1 % of the travelled distance. Despite superior performance compared to hand-held devices, shoe mounted IMUs are considered as less practical for consumers, because such systems require a special arrangement to integrate an IMU into a shoe.

Nilsson et al. (2010) have developed a methodology for performance assessment of foot-mounted and ZUPT assisted IMUs. As a result from variance propagation, the achievable performance for such a system is in the range of 0.1 % to 0.44 %, depending on external influences.

14.4 Foot Mounted Pedestrian Navigation

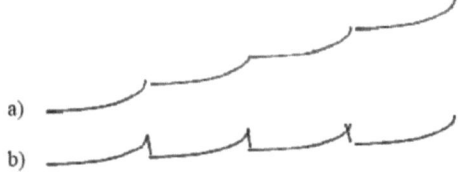

Figure 14.3 Drift of foot mounted IMUs according to (Wan and Foxlin 2010), using a) ZUPT, b) EKF zero-velocity pseudo measurements

Skog et al. (2010) studied the performance of different zero-velocity detectors for foot mounted inertial navigation. Four different strategies of zero-velocity updates have been investigated, namely the acceleration moving variance detector, the acceleration magnitude detector, the angular rate energy detector, and a new development, referred to as SHOE detector. The last two zero-velocity updates showed the highest position accuracy with 0.14 % deviation of the total travelled distance.

InterSense Inc. is currently prototyping a foot mounted inertial navigation system named NavShoe™ for the purpose of guiding first responders in hazardous environments. The position data from a dead reckoning module can be fused with GNSS or alternatively with RF observations. The unassisted system of Wan and Foxlin (2010) reached a drift deviation of 0.27 % of the travelled distance.

14.4.1 Combination of Foot Mounted INS and Maps

As with the general application for navigational IMUs in combination with maps (see 14.3.6), foot mounted inertial data can be fused with map matching. A pedestrian guidance system for assisting rescue teams has been suggested by Walder and Bernoulli (2010). The approach uses ZUPT where the constraints from a semantic Building Information Model (BIM) are assumed to be available. The positioning accuracy depends on the properties of the BIM and the quality of the map-matching algorithm.

14.4.2 Combination of Foot Mounted INS and Signal Strength

The foot mounted system proposed by Jiménez et al. (2010) relies on Received Signal Strength (RSS) from active RFID tags placed at known coordinates in a building. At an emission frequency of 1 Hz, the battery life time is expected to support the INS based pedestrian navigation system for 6 months. In order to take advantage of ZUPT, the INS sensor is attached to the foot. In addition, Zero Angular Rate Updates (ZARU) during stance stages of the walk are used. From the self-contained information alone a deviation of 1 % of the travelled distance is reported. As with RSSI from RFID, signal strength from WLAN can be combined with foot mounted INS. Frank et al. (2009) reduced the mean position deviation from 3.2 m to 1.6 m by fusion of WLAN fingerprinting and shoe data.

14.4.3 INS Mounted on Both Feet

Placing IMUs on both feet provides the ZUPT approach with further reliability, redundancy and continuity of service. Mounting a MEMS inertial unit to each foot of a pedestrian provides two independent navigation solutions. During the human gait cycle the separating distances between these two position solutions can be used as constraints. Bancroft et al. (2008) integrated the inertial data of two IMUs into a single Kalman filter and compared the single IMU approach and the twin approach with each other. Test runs showed that the position accuracy could be improved by more than 60 %.

14.5 Summary on INS Based Systems

The performance of INS based pedestrian navigation depends on several factors: the frequency of absolute positioning updates provided by a complimentary sensor system, the quality of the IMU used and the amount of additional information derived from the human's gait cycle. Foot-mounted systems can make use of zero velocity during the foot is in stance stage and have therefore a drift of less than 1 % of the travelled distance compared to IMUs mounted at other body parts with drifts being typically larger than 1 %. INS based navigation is the only approach which can be used without any infrastructure – at least for a certain time span. A comparison of INS based approaches included in this survey is given in Table 14.1.

Table 14.1 Pedestrian navigation approaches, sensors used and performance parameters as reported

Name	Mounting Body Part or Device	Accuracy (% of Distance Travelled)	Accelerometer	Gyroscopes	Barometer	Magnetomet.	GNSS	Map Info.	RSSI	ZUPT	ZARU	Local Reference	IMU Sensors	EKF	Maturity
Lukianto	phone	-	3	3	y	y	y	-	-	-	-	WLAN, Bluetooth	MTi-G Xsens	y	suggest.
Mathiassen	unspecified	23 %, 10 m / 60 s	3	3	y	3	y	-	-	-	-	GNSS, Barom.	ADIS16405	y	study
Sun	belt	2%	3	3	-	y	y	-	-	-	-	GNSS	AK8976A	y	study
Beauregard	helmet	few %	3	3	-	3	y	-	-	-	-	GNSS	MT9 Xsens	-	study
Bao	limbs	activity detection	9	-	-	-	-	-	-	-	-	-	ADXL201	-	study
Shin	phone	1% - 5%	3	3	-	3	-	-	-	-	-	-	MEMS	-	study
Kemppi	waist/pocket	16%, 17 m	3	3	-	-	-	-	-	-	-	map, beacon	Accelerometer	PF	study
Aufderheide	camera	-	3	3	-	3	-	-	-	-	-	images	MEMS	y	suggest.
Keßler	chest / foot	-	3	3	y	y	-	y	-	y	-	images	SCA / ADIS16255	y	study
Liu T.	backpack	1 %	3	3	-	3	-	-	-	-	-	scanners, Images	InertiaCube3	-	study
Seitz	phone	5 m	3	-	-	3	y	-	y	-	-	WLAN RSSI	Bosch BMA150	-	study
Fink	belt	0.67 m	2	1	-	-	-	-	-	y	-	RSSI	LIS3LV	y	study
Schmid	body	5 %, 3 m	3	3	-	-	-	-	y	-	-	ZigBee RSSI	MTi-G Xsens	y	study
Klingbeil	waist	1 m – 6 m	3	3	y	3	y	y	-	-	-	GPS, US,RF, CSS	Accelrometer	y	prod
Gusenbauer	phone	4%	3	-	-	y	y	y	-	-	-	map, A-GNSS	N97	y	study
Wagner	phone/car	1 – 15 m	y	y	-	-	y	y	-	-	-	map	built in CAR	y	test
Pressl	Belt	1 – 2 m	3	1	y	3	y	y	-	-	-	map, GNSS	PNM Vectronix	y	prototype
Nilsson	foot	0.1% – 0.44%	3	3	-	-	-	-	-	y	-	-	-	KF	simulation
Skog	foot	0.14 – 0.20%	3	3	-	-	-	-	-	y	y	-	MicroStrain3DM	y	study
Wan NavShoe	foot	0.14 %	3	3	-	-	-	-	-	y	-	test without	NavChip	y	prototype
Walder	foot	≈building model	3	3	y	y	y	-	y	-	-	map matching	MTi-G Xsens	y	study
Jiménez	foot	1% - 1 m	3	-	3	-	-	y	y	y	-	RFID RSSI	MTi-G Xsens	y	study
Frank	foot	1.6 m	3	3	-	y	-	-	y	y	-	WLAN RSSI	MEMS	y	study

15 Magnetic Localization

Wavelength	100 nm	1 μm	10 μm	100 μm	1 mm	10 mm	0.1 m	1 m	10 m	100 m	1 km
	UV		Infrared			Microwave			Radio		
Frequency	3 PHz	300 THz	30 THz	3 THz	300 GHz	30 GHz	3 GHz	300 MHz	30 MHz	3 MHz	300 kHz

Positioning systems using artificial magnetic and electromagnetic fields are described in this chapter. Magnetic fields can be generated from permanent magnets or from coils using Alternating Current (AC) or pulsed Direct Current (DC) fields. Electromagnetic fields can also be used for positioning in combined use of their electric field and magnetic field. The two sources of electromagnetic fields are static charges producing electric fields and currents producing magnetic fields. Oscillating charges produce electric and magnetic fields.

15.1 Systems Using the Antenna Near Field

The Near-Field Electromagnetic Ranging (NFER) uses the properties of radio waves, where the near field encompasses an antenna or, more generally, any electromagnetic radiation source with an approximate sphere of radius 1/6 of the radiation wavelength (Capps 2001). In NFER, the distance from a small transmitter antenna is derived from the phase relation between the electric and the magnetic field components of an electromagnetic field. The receiver unit must be able to receive the two signal components separately and compare their phases. Close to the antenna, these components have a maximal phase difference of 90°. Since the phase difference decreases with the distance to the antenna it can be used for range determination within a certain proximity to the antenna. As a major advantage NFER does not require synchronization or signal modulation. Secondly, if low frequencies around 1 MHz are used, the signals have the potential to penetrate walls. On the other hand, the use of low RF frequencies requires large receiver units since an efficient receiver antenna needs to be within a quarter-wavelength in size.

The 2D location system Q-Track, characterized in Schantz et al. (2011) makes use of the NFER principle. The system is designed for an operating range of $\lambda \cdot (2\pi)^{-1}$ where the applied wavelength is $\lambda = 300$ m (1 MHz). Measurements are taken in an office environment with non-line-of-sight conditions (i.e. through the walls). The reported average distance deviation is 55 cm with an operating range between 1.4 m and 23 m.

An accuracy of 1 m is reported for 83 % of the *x*- and *y*-positions which are determined from multilateration using 5 fixed receivers.

15.2 Systems Using Magnetic Fields from Currents

Magnetic fields are produced by magnetic material or electrical currents. A type of positioning system makes exclusive use of the strength and the direction of the magnetic field. Two related, but separately treated measures exist for the magnetic field: the magnetic field strength H (A/m) and the magnetic flux density B (Tesla). The relation of the two quantities is

$$B = \mu H, \tag{15.1}$$

where μ is the a material dependent parameter known as permeability. The permeability varies not only with the conductivity of the material but also with the temperature and the frequency of the field. Despite the variability of μ, providers of magnetic positioning systems use the simplified term 'magnetic field' considering H and B to be proportional.

Direct Current (DC) magnetics use pulsed direct currents where the current frequency is low enough to be considered static. Static magnetic fields are caused by different direct current sources, such as coils or wires.

15.2.1 Systems Using Coils

An artificial quasi static magnetic field can be created by electrical coils. A useful property of a coil based magnetic field is that it can be well predicted from a theoretic model. The flux density of a vertically orientated coil is

$$B(\phi,r) = \frac{\mu_0 u I F}{4\pi r^4}\sqrt{1+3\sin^2\phi}, \tag{15.2}$$

where μ_0 is permeability of vacuum, u the number of loops, I the electric current, F the area of the coil, ϕ the elevation angle of a point **P** in relation to the coil and r the distance of that point to the coil. Due to rotational symmetry of a coil, the flux density B is independent of the horizontal angle. At a mobile magnetic field sensor, B can be measured at any point **P** within the field. Equation (15.2) can be resolved for distance r and elevation angle ϕ. If multiple coils generate magnetic fields at known locations, the relative 3D position of the sensor can be determined. The coil-based approach has several advantages: a) it can be used under NLoS conditions, b) the magnetic field is not affected by reflection or multipath and c) kinematic tracking is possible at high measurement rates with an unlimited number of sensors.

Blankenbach and Norrdine (2010) have built an experimental model using a coil with 0.5 m in diameter and 140 loops. A ranging accuracy of a few centimeters was achieved for short distances of less than 10 m. For larger distances, the magnetic field turned out to be too weak and therefore vulnerable to noise. One of the results was that there is a discrepancy between theoretical model and observation, which indicates a requirement for prior system calibration.

Ascension (2011) provides the coil based system 'track STAR' which uses pulsed DC magnetic field transmitters with operating ranges between 0.8 m and 4.2 m. For the version with large area coverage of 4 m, the positioning accuracy of static objects is stated as 3.8 mm.

Arumuam et al. (2011) have used an emitter driven at 387 kHz consisting of 45 turns of a copper wire to generate a magentoquasistatic field. Theoretical results derived from infinitesimal dipole approximation are compared with experimental results, indicating that distances up to 50 m can be estimated with an accuracy of 20 cm.

15.2.2 Systems Using AC Magnetics

Unlike systems based on coils using pulsed DC technology, AC (Alternating Current) based magnetic tracking systems are less affected by the Earth's magnetic field and artificial magnet fields from electric devices.

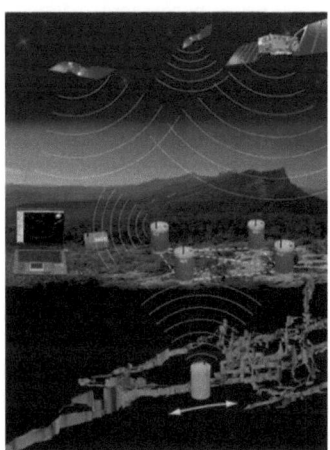

Figure 15.1 Principle of 'Underground GPS', InfraSurvey (2011)

The electromagnetic tracker system LIBERTY from Polhemus (2011) uses AC magnetic fields and up to 16 sensors to observe position and orientation of an object at update rates of 240 Hz. According to the reported sub-millimeter accuracy, medical applications are possible within an operating range of 1.5 m which can be extended up to 4.6 m.

InfraSurvey (2011) is offering the underground positioning system UGPS based on low frequency AC magnetic fields. It has been designed for measuring position and orientation of objects in underground environments such as mines, tunnels, caves or pipes, but can also be used for indoor positioning. Based on distance measurements between static receivers deployed at the surface, it is possible to locate an active magnetic transmitter weighing 8 kg within 1 m precision at a distance of 100 m and a maximum operating distance of 200 m. The position of the above-surface receiver stations are determined by GNSS. The principle of UGPS is illustrated in Figure 15.1.

15.3 Systems Using Permanent Magnets

The second method of using the magnetic flux density for positioning is through magnetic fields created by permanent magnets. A typical system consists of multiple static magnetic sensors which measure the magnetic flux density of a mobile magnet. Alternatively, multiple static permanent magnets at known locations can be used to locate a mobile magnetic field sensor. Current approaches indicate a measurement volume of 1 m³, which restricts the method to close range usage, such as medical applications. A major challenge in positioning using permanent magnets arises from the complexity of the magnetic field. Mathematic models involve high order nonlinear equations which create a multimodal objective function with multiple local optima for the position solution.

Song et al. (2009) locate a cylindrical permanent magnet from 64 deployed magnetic sensors within a space of 0.5 m². At a measurement rate of 67 Hz and a latency of 15 ms, they achieve an average position deviation of 1.8 mm and an angular deviation of 1.5° in the magnet's orientation. The system addresses the application of 3D positioning of a capsule endoscope in a human body.

Liu W. et al. (2010) combine optical tracking and magnetic localization in order to overcome the occlusion problem that optical systems face. A permanent magnet is tracked by a dense array of sensors which measure the magnetic flux intensity in three dimensions within a cubic shaped magnetic field. The advantage of using a magnetic system component is that line of sight between the magnetic sensor and the tracked object is not a requirement. The optical module is a redundant system component which consists of 4 video cameras and is used to enhance the robustness to disturbance by ferromagnetic objects within the working volume. The reported positional accuracy for the combined system is 1 mm within a volume of 1 m³.

15.4 Systems Using Magnetic Fingerprinting

The idea of magnetic fingerprinting arises from animals that determine their position from local anomalies of the Earth's magnetic field. Likewise in buildings, each location has its unique signature of its magnetic flux density. These fluctuations in space arise from natural and man-made sources, such as metal building material, electric power systems and industrial devices. The anomalies of the magnetic field have sufficient variability in space to be detected by a magnetometer. Under the assumption that the magnetic field inside a building is approximately static, a fingerprinting method can be applied. A magnetic map of the rooms is taken in a setup phase, and the current location is determined by comparing the current flux density with the flux density values stored in the database.

Haverinen and Kemppainen (2009) have mounted a 3-axis magnetometer on a robot to determine its location within a building by magnetic fingerprinting. The magnetometer has been mounted at the end of a rod with a length of 0.4 m to avoid influence of the ambient magnetic field from the robot's motor. Subsequent to a calibration phase, the detection of the robot's location along a path length of 278 m was possible. The robot needed to travel 25 m on average in order to get localized by comparing the magnetic flux values. Along that one-dimensional path, the reported accuracy was 0.2 m.

15.5 Summary on Magnetic Localization

Unlike other technologies, magnetic localization does not require the maintenance of a line-of-sight between sensor and source. Therefore, the use of electric and magnetic velocity fields is advantageous if walls need to be penetrated and is the only way to detect structures buried deep underground. Different approaches range from systems dedicated for surgery with less than 1 m^3 volume operating at mm-accuracy level up to large coverage, low accuracy fingerprinting systems, see Table 15.1 for a comparison.

Table 15.1. Approaches based on magnetic localization techniques and reported performance parameters

Name	Year	Principle	Coverage Area	Application	Reported Accuracy	Market Maturity
Q-Track	2011	near field	23 m	NLoS office & industry	50 cm	experimental system
Blankenbach	2010	DC field, coils	10 m	NLoS indoor	few cm	experimental system
Ascension	2011	DC field, coils	4.2 m	medical instr. guidance	1-4 mm	commercial system
Arumugam	2011	DC field, coils	50 m	American football	20 cm	experimental system
InfraSurvey	2011	AC magnetic field	200 m	caves, mines, tunnels	1 m	commercial system
Polhemus	2011	AC magnetic field	1.5 m (5 m)	head & body tracking	0.7 mm	commercial system
Song	2009	permanent magnet	0.5 m	endoscope	2 mm	experimental system
Liu	2010	permanent magnet	1 m	assisted surgery	1 mm	experimental system
Haverinen	2009	fingerprinting	280 m (1D)	robot localization	20 cm	experimental system

16 Infrastructure Systems

This chapter outlines indoor positioning approaches which cannot be matched to any of the technologies discussed in the previous chapters. These technologies make use of existing building infrastructure or embed additional infrastructure into the building material. The developed systems can be hidden from its users into the structures of the building.

16.1 Power Lines

Power Line Positioning (PLP) is a fingerprinting-based method to provide sub-room positioning in a household based on the existing electrical grid. In comparison to RFID systems which require dense deployment of RFID chips, PLP simply uses the power line infrastructure in a building. The principle of PLP is that unmodulated carrier wave signals in the frequency range between a few kHz to 20 MHz are generated by an interface module plugged into an electrical outlet in a home. These signals consist of energy rich electrical transients and can be wirelessly received by passive tags. At the tag's resonance frequency, the resonator inductively couples back a signal into the power line. This signal can be detected in a so-called power line interface as a decaying swinging. If multiple tags with differing resonance frequencies are used, an individual identification of the tags becomes possible.

Patel et al. (2009) present a proof of concept of their power line location system based on battery-less tags. For the current system the read distance along the power line is 3 m to 4 m and the maximal reading distance between the tags and the electrical wiring is 50 cm. At this stage, the system only provides the detection of the tags rather than an actual location which can be associated with a positioning accuracy. Despite these limitations Patel et al. see applications as finding lost items and the detection of the absence of a tagged item. Stuntebeck et al. (2008) have employed wideband signals with frequencies between 447 kHz and 20 MHz and taken 66 fingerprints in a 0.9 m by 0.9 m grid. Their results show that it is possible to achieve grid-level accuracy. A drawback of the system is its low temporal stability and the degradation of the system in environments with disturbing electrical equipment. As a fingerprinting-based technology, there is also the requirement of an initial site survey which includes signal amplitude measurements at several fingerprints locations.

16.2 Floor Tiles

Passive, unobtrusive indoor tracking of human beings can be achieved using multiple floor tiles. The main advantage is that they are invisible to the user and that the users do not need to be equipped with any tags. Such systems can detect a standing human in 2D up to an accuracy of 1 dm, allowing for applications in healthcare and AAL. There are different sensor techniques that build on the concept of smart carpets.

The tile track system of Valtonen et al. (2009) is based on measuring the capacitance between multiple floor tiles. When a person is standing over a tile, the capacitance between the human feet and a transmitter increases. The system can track multiple persons, if they stay at least one tile apart from each other. Valtonen et al. report an accuracy of 15 cm for a standing human and 41 cm for a walking person. A large-area capacitive system designed for Ambient Assisted Living (AAL) has been described in Steinhage et al. (2007) and commercialized under the name SensFloor (2011). The individual mats of the textile underlay system SensFloor deliver the acquired data to a central unit via a radio link. Rimminen et al. (2009 and 2010) present a floor sensor system, which uses the electrical near-field for fall-detection.

Another way of building a tile tracking system is to measure the pressure under humans' feet. Arrays of force-sensitive resistors can detect the pressure on tiles deployed on the floor. Richardson et al. (2004) describe the pressure-based floor track system 'Z-Tiles' that determines the position through a series of hexagonal tiles, which join together to form a flexible pressure sensing surface. In order to forward the pressure information, a self-organized network is set up. Richardson et al. report a resolution of 4 cm and a response time of a few milliseconds for their system which has been primarily designed for interactive dance floors and gaming.

16.3 Fluorescent Lamps

The principle of optical communication can be used to employ lamps for the purpose of indoor guidance. It is possible to employ fluorescent light tubes which have become a common type of light source in office buildings. Free-space optical data transmission using electromagnetic waves in the visible or infrared bands is enabled by modulation of the 'lamp current'. A switch-mode control device produces a variation of phase or frequency of the lamp's AC current, creating a modulation of the produced luminous flux and generating a signal with a code being unique to every lamp. The modulation frequency is chosen to be greater than 20 kHz, such that audible noise is avoided. As a light sensor on a mobile platform passes the modulated light source, its position is determined by reading the unique code of the nearest lamp.

Liu et al. (2006) have demonstrated feasibility for an implementation of an indoor guidance system for the blind using fluorescent lights. A positioning system based on light communication has been combined with an IMU based navigation system by Nishikata et al. (2011). The approach uses visible light signals to provide absolute position updates for a dead reckoning system.

16.4 Leaky Feeder Cables

A Leaky feeder consists of a long coaxial cable deployed along corridors to provide radio services in buildings and underground environments. As its name implies, a leaky feeder is designated to let radio signals leak out of the cable along its length. While this technology was originally designed for communication, existing infrastructure of leaky feeder cables can be used for indoor positioning based on radiated RF signals. Although homogeneous radio signal coverage throughout the cable is wanted for communication, there is an unavoidable attenuation along its entire length. The loss in signal strength can be used to create a RSSI fingerprinting map which shows a distinct relation between location and signal amplitude. The position of a receiver can be deduced during an online phase via comparison of the current fingerprint and those in the database. The advantage of using RSSI of leaky feeders is the robustness to environmental changes. Radio technologies used for leaky feeder cables are GSM and WLAN.

Weber et al. (2011) carried out experiments with a 40 m WLAN leaky feeder deployed on the ceiling of an office hallway. Based on 41 fingerprints with a separation of 1 m, it was possible to determine the 1D position with an overall accuracy of 4 m.

Pereira et al. (2011) carried out similar studies on an existing leaky feeder cable installed throughout the tunnel for the LHC accelerometer in Switzerland. The leaky feeder supplies the whole tunnel of 27 km perimeter with GSM signal coverage. Fingerprints were taken in intervals of 200 m along the entire tunnel where the longitudinal attenuation (with factors of about 4 dB/km) could be related to the tunnel chainage within an accuracy of ca. 200 m. This result could be improved by an increased resolution of the calibration map.

16.5 Summary on Infrastructure Systems

Existing building infrastructure as well as infrastructure deployed unobtrusively in the building interior for the purpose of indoor positioning can be successfully used for providing positioning services within a building – at various levels of accuracy and costs. In Table 16.1 performance parameters of different infrastructure based systems are quantified.

16.5 Summary on Infrastructure Systems

Table 16.1. Infrastructure based systems

Name	Year	Principle	Existing or Deployed Infrastructure	Coverage	Application	Reported Accuracy	Market Maturity
Patel	2009	power lines	existing	3 m - 4 m	finding lost items	detection	study
Stuntebeck	2008	power lines	existing	building	location aware homes	1 m – 3 m	study
Valtronen	2009	floor tiles	deployed	2.4 × 2.0 m	assistance for elderly	15 cm-40 cm	prototype
SensFloor	2011	floor tiles	deployed	50 m^2	assistance for elderly	dm	product
Rimminen	2009	floor tiles	deployed	19 m^2	fall detection	21 cm	prototype
Richardson	2004	floor tiles	deployed	< 4 m^2	dance floor	> 4 cm	prototype
Liu	2006	fluorescent lamps	existing	building	guidance for the blind	-	study
Nishikata	2011	fluorescent lamps	existing	building	robot guidance	10 cm	study
Weber	2011	leaky feeder	deployed	40 m	indoor localization	4 m	study
Pereira	2011	leaky feeder	existing	27 km	unmanned processing	200 m	study

17 Concluding Remarks

17.1 Conclusion

The diversity of different technological solutions for indoor positioning and navigation shows how profoundly interdisciplinary the field is and reflects that almost any signal/sensor technique can be exploited for this purpose. Despite the abundance of approaches which exist to tackle the indoor positioning problem, current solutions cannot cope with the performance level that significant applications require. In short, requirements for the mass market include 1 m horizontal accuracy, floor identification, absence of coverage-gaps, >99 % availability and minimal costs for local installations. Apart from insufficiency in position accuracy, coverage and availability, the need for extensive node deployment and maintenance is the main reason why system implementations are not sufficiently economical. A good fraction of research approaches are also missing appealing usability to enable wide-scale consumer adoption.

17.2 Outlook

To improve this situation of insufficiency in performance, two tasks need to be performed. First, user requirements need to refined, i.e. specific determination and quantification of requirements parameters for every application. These figures provide essential guidelines for future focus in research and implementation of efficient indoor positioning systems. The second task is thorough performance benchmarking of implemented systems. In this context, benchmarking should not be understood as simple comparison of different indoor positioning systems, but also as the task of finding the optimal match between quantified requirements and assessed performance parameters for each application separately. Fulfillment of the two tasks is not straightforward due to a large number of different performance criteria which need to be weighed against each other in the form of a priority list. Completion of benchmarking was out of scope for this work. In order to be successful, users, providers, developers and manufacturers of indoor positioning systems must collaboratively discuss potential achievements for each application. While solutions can be found for specific indoor positioning tasks with a limited area of influence, an overall solution – such as GNSS for outdoor environments – can only be found after agreement on standards in communication and protocols.

Acronyms

Acronym	Expansion / Meaning
2D	Two Dimensions
3D	Three Dimensions
AAL	Ambient Assistant Living
AC	Alternating Current
AGNSS	Assisted GNSS
AGPS	Assisted GPS
AoA	Angle of Arrival
AP	Access Point
AR	Augmented Reality
ATR	Automatic Target Recognition
BIM	Building Information Model
BN	Blind Node
BS	Base Station
CAD	Computer-Aided Design
CCD	Charge Coupled Device
CDMA	Code Division Multiple Access
CIR	Channel Impulse Response
CityGML	City Geography Markup Language, a common information model for the representation of 3D urban objects
CMOS	Complementary Metal Oxide Semiconductor
CRLB	Cramér-Rao Lower Bound
CSS	Chirp Spread Spectrum
dB	Decibel, (= 0.1 Bel)
dBm	Power ratio in decibels (dB) of the measured power referenced to one milliwatt
dBW	Decibel Watt, a unit for the signal strength expressed in decibels referenced to one watt
DC	Direct Current
DECT	Digital Enhanced Cordless Technology
DGNSS	Differential Global Navigation Satellite System
DoF	Degrees of Freedom
DPM	Dominant Path Model
DR	Dead Reckoning
ECC	European Communications Committee

ECoO	Enhanced Cell of Origin
EDM	Electronic Distance Meter
EKF	Extended Kalman Filter
EPAM	Extended Phase Accordance Method
FCC	Federal Communications Commission
FM	Frequency Modulation
FMCLR	Frequency Modulated Coherent Laser Radar
FMCW	Frequency Modulated Continuous Wave
FP	Fingerprinting
GIS	Geographic Information System
GLONASS	Globalnaja Nawigazionnaja Sputnikowaja Sistema, a GNSS system
GNSS	Global Navigation Satellite System, any of the existing or proposed satellite-based positioning systems, such as GPS, GLONAS, Galileo and Beidou
gon	Unit where a circle measures to 400 degrees
GPS	Global Positioning System, a GNSS system
GSM	Global System for Mobile communication
HF	High Frequency (3 MHz to 30 MHz)
HSGNSS	High Sensitive Global Navigation Satellite System
Hz	Hertz, SI name for cycles per second
ID	Identification (number)
IEEE	Institute of Electrical and Electronics Engineers
iGPS	iGPS (indoor Global Positioning System) a laser-based 3D measurement system offered by Nikon Metrology
IMU	Inertial Measurement Unit
INS	Inertial Navigation System
IR	Infrared
ISM	Industrial, Scientific and Medical radio band, reserved internationally for the use of radio frequency other than communications
JCGM	Joint Committee for Guides in Metrology
KF	Kalman Filter
KNN	K-Nearest Neighbor
LAN	Local Area Network
LBS	Location Based Services
LED	Light Emitting Diode
LF	Low Frequency (30 kHz - 500 kHz)
LHC	Large Hadron Collider, a particle accelerator near Geneva
LoD	Level of Detail
LoS	Line of Sight
LPS	Local Positioning System
MAC	Media Access Control (layer)
MEMS	Micro Electro-Mechanical System
MDS	Multi Dimensional Scaling
MIMO	Multiple-Input Multiple-Output
MLS	Maximum Length Sequences
MM	Map Matching
MS	Mobile Station
MT	Mobile Terminal
MWM	Multi Wall Model
NFER	Near Field Electromagnetic Ranging

NLI	Natural Language Instructions
NLoS	Non Line of Sight
PD	Phase Difference
PDR	Pedestrian Dead Reckoning
PLP	Power Line Positioning
PoA	Phase of Arrival
PRN	Pseudo Random (sequence) Number
PSD	Position Sensitive Device
RADAR	RAdio Detection And Ranging
RF	Radio Frequency
RFC	Receptive Field Coocurrence
RFID	Radio Frequency IDentification
RMSD	Root Mean Square Deviation
RR	Response Rate
RSS	Received Signal Strength
RSSI	Received Signal Strength Indicator
RTLS	Real Time Locating System
RToF	Roundtrip Time of Flight
RTT	Round Trip Time
SDS-TWR	Symmetrical Double-Sided Two Way Ranging
SF-CW	Stepped Frequency Continuous Wave
SHF	Super High Frequency (3 GHz - 30 GHz)
SLAM	Simultaneous Localization And Mapping
SNR	Signal to Noise Ratio
SoL	Safety of Life
TDoA	Time Difference of Arrival
ToA	Time of Arrival
ToF	Time of Flight
TTFF	Time To First Fix
TWR	Two Way Ranging
UHF	Ultra High Frequency (300 MHz - 3 GHz)
US	Ultra Sound
UTMS	Universal Mobile Telecommunications System
UWB	Ultra-Wideband
VHF	Very High Frequency (30 MHz to 300 MHz)
WLAN	Wireless Local Area Network
WPAN	Wireless Personal Area Network
WSN	Wireless Sensor Networks
ZARU	Zero-Angular Rate Update
ZUPT	Zero velocity-UPdaTe

Symbols

Symbol	Expansion / Meaning	Symbol	Expansion / Meaning
A	amplitude	m	number of fingerprints
$\hat{a}_P$	estimated variance of a point P	n	number of points / nodes
a_P	average position deviation of point P	n_m	number of mobile positions
α	averaged fast fading term	n_r	number of ranges
B	magnetic flux density	n_s	number of static nodes
b	bandwidth	μ	permeability
c	empirical pass loss penetration factor	μ_0	permeability of vacuum
$\mathbf{c}$	RSSI calibration vector	$\mathbf{P}$	position vector
$\mathbf{C_x}$	variance-covariance matrix	$\hat{\mathbf{P}}$	estimated position
c_0	speed of light in vacuum	p	path loss exponent
cor	correlation coefficient	P_R	received power
d	distance	P_T	transmitted power
$\mathbf{d}$	vector of distances	θ	heading
Δf	frequency spectrum	ϕ	elevation angle
Δt	time interval	q	number of joints
F	area of a coil	r	range
f	radio frequency	rr	range resolution
$\mathbf{f}$	fingerprint	$r(t)$	multipath signal
G_R	antenna gain of receiver	$\mathbf{r}_t$	RSSI received at time t
G_T	antenna gain of transmitter	$s(t)$	transmitted signal
γ	model parameter for slow fading	σ	standard deviation
H	magnetic field strength	σ_P	standard deviation of position P
h	model parameter for fast fading	τ	delay of a propagation path
I	electric current	T	temperature
i	point / node number	t	time
j	fingerprint number	u	number of loops of a coil
k	number of floors	v	velocity, speed
L	total pass loss	x, y, z	coordinates
l	step length	$\mathbf{x}$	state vector
λ	wavelength	$z(t)$	random noise

References

Aalto, L., Gothlin, N., Korhonen, J. and Ojala, T. (2004): "Bluetooth and WAP Push Based Location-aware Mobile Advertising System", Proceedings of the 2nd International Conference on Mobile Systems, Applications, and Services, MobiSYS '04, pp. 49–58.

ABATEC (2012): http://www.abatec-ag.com/, last accessed 11. October 2011.

AICON 3D Systems (2011): http://www.aicon.de, last accessed 14. October 2011.

Alawieh, M., Patino-Studencka, L. and Dahlhaus, D. (2010): "Stochastic Modeling of Pseudolite Clock Errors Using Enhanced AR Methods", Proceedings of CSNDSP 2010, pp. 178–183.

Alloulah, M. and Hazas, M. (2010): "An Efficient CDMA Core for Indoor Acoustic Position Sensing", Proceedings of the 2010 International Conference on Indoor Positioning and Indoor Navigation (IPIN), September 15–17, 2010 Campus Science City, ETH Zurich, Switzerland.

Ambiplex (2011): http://www.ambiplex.com/, last accessed 14. October 2011.

Arumugam, D., Griffin, J., Stancil, D. and Ricketts, D. (2011): "Higher Order Loop Corrections for Short Range Magnetoquasistatic Position Tracking", Proceedings of the 2011 IEEE Antennas and Propagation Conference (APURSI), July 2011, pp. 1755–1757.

Ascension (2011): http://www.ascension-tech.com/, last accessed 17. May 2011.

Atsuumi, K. and Sano, M. (2010): "Indoor IR Azimuth Sensor Using a Linear Polarizer", Proceedings of the 2010 International Conference on Indoor Positioning and Indoor Navigation (IPIN), September 15–17, 2010 Campus Science City, ETH Zurich, Switzerland.

Aufderheide, D. and Krybus, W. (2010): "Towards Real-Time Camera Egomotion Estimation and Three-Dimensional Scene Acquisition from Monocular Image Streams", Proceedings of the 2010 International Conference on Indoor Positioning and Indoor Navigation (IPIN), September 15–17, 2010 Campus Science City, ETH Zurich, Switzerland.

References

Bahl, P. and Padmanabhan, V. (2000): "Radar: An In-Building RF-based User Location and Tracking System", Proceedings of INFOCOM 2000, IEEE Conference on Computer Communications, Tel Aviv, Israel.

Bancroft, J., Lachapelle, G., Cannon, M. and Petovello, M. (2008): "Twin IMU-HSGPS Integration for Pedestrian Navigation", Proceedings of the 21st International Technical Meeting of the Satellite Division of The Institute of Navigation (ION GNSS 2008), pp. 1377–1387.

Bao, L. and Intille, S. (2004): "Activity Recognition from User-Annotated Acceleration Data", Proceedings of the 2nd International Conference on Pervasive Computing, pp. 1–17.

Bargh, M. and de Groote, R. (2008): "Indoor Localization Based on Response Rate of Bluetooth Inquiries", ACM 2008, pp. 49–54.

Barnes, J., Rizos, C., Kanli, M., Pahwa, A., Small, D., Voigt, G., Gambale N. and Lamance, J. (2005): "High Accuracy Positioning Using Locata's Next Generation Technology", Proceedings of the 18th International Technical Meeting of The Institute of Navigation (ION GNSS 2005), pp. 2049–2056.

Barnes, J., Van Cranenbroeck, J., Rizos, C., Pahwa, A. and Politi, N. (2007): "Long Term Performance Analysis of a New Ground-transceiver Positioning Network (LocataNet) for Structural Deformation Monitoring Applications", FIG Working Week 2007, Hong Kong SAR, China, 13–17 May 2007.

Baum, M., Niemann, B., Abelbeck, F., Fricke, D.-H. and Overmeyer, L. (2007): "Qualification Tests of HF RFID Foil Transponders for a Vehicle Guidance System", Proceedings of the 2007 IEEE Intelligent Transportation Systems Conference (ITSC 2007), pp. 950–955.

Beauregard, S. (2006): "A Helmet-Mounted Pedestrian Dead Reckoning System," Proceedings of the 3rd International Forum on Applied Wearable Computing (IFAWC 2006), pp. 15–16.

Bensky A., (2007): "Wireless Positioning Technologies and Applications", Artech House Publishers 2007, 311 p.

Blankenbach, J. and Norrdine, A. (2010): "Position Estimation Using Artificial Generated Magnetic Fields", Proceedings of the 2010 International Conference on Indoor Positioning and Indoor Navigation (IPIN), September 15–17, 2010 Campus Science City, ETH Zurich, Switzerland.

Bolliger, P. (2008): "Redpin – Adaptive, Zero-Configuration Indoor Localization Through User Collaboration," Proceedings of the ACM International Workshop on Mobile Entity Localization and Tracking in GPS-less Environments, pp. 55–60.

Boochs, F., Schütze, R., Simon, C., Marzani, F., Wirth, H. and Meier, J. (2010): "Increasing the Accuracy of Untaught Robot Positions by Means of a Multi-Camera System", Proceedings of the 2010 International Conference on Indoor Positioning and Indoor Navigation (IPIN), September 15–17, 2010 Campus Science City, ETH Zurich, Switzerland.

Bürki, B., Guillaume, S., Sorber, P. and Oesch, H. (2010): "DAEDALUS: A Versatile Usable Digital Clip-on Measuring System for Total Stations", Proceedings of the 2010 International Conference on Indoor Positioning and Indoor Navigation (IPIN), September 15–17, 2010 Campus Science City, ETH Zurich, Switzerland.

Capps, C. (2001): "Near Field or Far Field", in Electronic Design News (EDN), August 2001, pp. 95–102.

Chen, Y.C., Chiang, J.R., Chu, H., Huang, P. and Tsui, A.W. (2005): "Sensor-Assisted Wi-Fi Indoor Location System for Adapting to Environmental Dynamics," Proceedings of ACM International Symposium on Modeling, Analysis and Simulation of Wireless and Mobile Systems, pp. 118–125.

Cheung, K., Intille, S. and Larson, K. (2006): "An Inexpensive Bluetooth-Based Indoor Positioning Hack", Proceedings of the 8th International Conference on Ubiquitous Computing, Extended Abstracts.

Chrysikos, T., Georgopoulos, G. and Kotsopoulos, S. (2009): "Site-Specific Validation of ITU Indoor Path Loss Model at 2.4 GHz", 10th IEEE International Symposium on a World of Wireless, Mobile and Multimedia Networks (WoWMoM), June 2009, pp. 1–6.

Ciurana, M., Lopez, D. and Barcelo-Arroyo, F. (2009): "SofTOA: Software Ranging for TOA-Based Positioning of WLAN Terminals", Proceedings of International Symposium of Location and Context Awareness '09, pp. 207–221.

Ciurana, M., Giustiniano, D., Neira, A., Barcelo-Arroyo, F. and Martin-Escalona, I. (2010): "Performance Stability of Software ToA-Based Ranging in WLAN", Proceedings of the 2010 International Conference on Indoor Positioning and Indoor Navigation (IPIN), September 15–17, 2010 Campus Science City, ETH Zurich, Switzerland.

Daly, D., Melia, T. and Baldwin, G. (2010): "Concrete Embedded RFID for Way-Point Positioning", Proceedings of the 2010 International Conference on Indoor Positioning and Indoor Navigation (IPIN), September 15–17, 2010 Campus Science City, ETH Zurich, Switzerland.

Depenthal, C. (2010): "Path Tracking with iGPS", Proceedings of the 2010 International Conference on Indoor Positioning and Indoor Navigation (IPIN), September 15–17, 2010 Campus Science City, ETH Zurich, Switzerland.

DeSouza, G. and Kak, A. (2002): "Vision for Mobile Robot Navigation: A Survey", IEEE Transaction on Pattern Analysis and Machine Intelligence, vol. 24, no. 2, pp. 237–267.

Diggelen, F. (2009): "A-GPS: Assisted GPS, GNSS, and SBAS", Artech House Publishers, 380 p.

Dziadak, K., Sommverville, K., Kumar, B. and Green, J. (2005): "RFID Applied to the Built Environment: Buried Asset Tagging and Tracking System", Proceedings of the 2nd Scottish Conference for Postgraduate Researchers of the Built and Natural Environment (PRoBE), 16–17 November 2005, pp. 493–504.

ECC Report-145 (2010): "Regulatory Framework for Global Navigation Satellite System (GNSS) Repeaters", Electronic Communications Committee (ECC), St. Petersburg, May

2010, http://www.erodocdb.dk/Docs/doc98/official/pdf/ECCREP145.PDF, last accessed 20. January 2011.

Eissfeller, B., Teuber, A. and Zucker, P. (2005): "Indoor-GPS: Ist der Satellitenempfang in Gebäuden möglich?", ZfV, vol. 4, pp. 226–234, 2005.

eRide (2011): www.eride.com, last accessed 15. November 2011.

EvAAL (2011): http://evaal.aaloa.org/, last accessed 20. November 2011.

Evolution Robotics (2010): http://www.evolution.com, last accessed 12. December 2010.

Filonenko, V., Cullen, C. and Carswell, J. (2010): "Investigating Ultrasonic Positioning on Mobile Phones", Proceedings of the 2010 International Conference on Indoor Positioning and Indoor Navigation (IPIN), September 15–17, 2010 Campus Science City, ETH Zurich, Switzerland.

Fink, A., Beikirch, H., Voß, M. and Schröder, C. (2010): "RSSI-based Indoor Positioning Using Diversity and Inertial Navigation", Proceedings of the 2010 International Conference on Indoor Positioning and Indoor Navigation (IPIN), September 15–17, 2010 Campus Science City, ETH Zurich, Switzerland.

Fischer, G., Klymenko, O., Martynenko, D. and Lüdiger, H. (2010): "An Impulse Radio UWB Transceiver with High-Precision TOA Measurement Unit", Proceedings of the 2010 International Conference on Indoor Positioning and Indoor Navigation (IPIN), September 15–17, 2010 Campus Science City, ETH Zurich, Switzerland.

Fluerasu, A., Jardak, N., Picois, A.V. and Smama, N. (2011): "Status of the GNSS Transmitter-Based Approach for Indoor Positioning", Coordinates – Positioning, Navigation and Beyond, vol. 7, no. 1, pp. 27–30.

Frank, K., Krach, B., Catterall, N. and Robertson, P. (2009): "Development and Evaluation of a Combined WLAN Inertial Indoor Pedestrian Positioning System", Proceedings of the 22nd International Technical Meeting of The Satellite Division of the Institute of Navigation (ION GNSS 2009), Savannah, Georgia, USA, pp. 538–546.

Frank, M. (2008): "Eine Idee nimmt Gestalt an 3D-Daten von physischen Modellen und Bauteilen erzeugen", Qualität und Zuverlässigkeit (QZ), vol. 53, no. 12, pp. 53–55.

Frick (2011): www.fricknet.com, last accessed 23. October 2011.

Fröhlich C. and Mettenleiter M. (2004): "Terrestrial Laser Scanning. New Perspective in 3D-Surveying", Proceedings of the ISPRS Working Group VIII/2, Freiburg, Germany, pp. 7–13.

Fujimoto, M., Nakamori, E., Inada, A., Oda, Y., Wada, T., Mutsuura, K. and Okada, H. (2011): "A Broad-Typed Multi-Sensing-Range Method for Indoor Position Estimation of Passive RFID Tags", Proceedings of the 2011 International Conference on Indoor Positioning and Indoor Navigation (IPIN), September 21–23, 2011 in Guimarães, Portugal.

References

Gallagher, T., Li, B., Kealy, A. and Dempster, A. (2009): "Trials of Commercial Wi-Fi Positioning Systems for Indoor and Urban Canyons", Proceedings of IGNSS Symposium, 2009, Gold Coast, Australia.

Gallagher, T., Li, B., Kealy, A., Dempster, A. and Rizos, C. (2010): "A Sector-Based Campus-Wide Indoor Positioning System", Proceedings of the 2010 International Conference on Indoor Positioning and Indoor Navigation (IPIN), September 15-17, 2010 Campus Science City, ETH Zurich, Switzerland.

Gansemer, S., Großmann, U. and Hakobyan, S. (2010): "RSSI-based Euclidean Distance Algorithm for Indoor Positioning Adapted for the Use in Dynamically Changing WLAN Environments and Multi-Level Buildings", Proceedings of the 2010 International Conference on Indoor Positioning and Indoor Navigation (IPIN), September 15-17, 2010 Campus Science City, ETH Zurich, Switzerland.

GATE (2011): http://www.gate-testbed.com/, last accessed 30. October 2011.

Golden, S. and Bateman, S. (2007): "Sensor Measurements for Wi-Fi Location with Emphasis on Time-of-Arrival Ranging", IEEE Transactions on Mobile Computing vol. 6, no. 10, pp. 1185-1198.

Grimm, D. (2012): "GNSS Antenna Orientation Based on Modification of Received Signal Strengths", PhD Dissertation at ETH Zürich, Institute for Geodesy and Photogrammetry, to be published.

Günther A. and Hoene, C. (2004): "Measuring Round Trip Times to Determine the Distance between WLAN Nodes", Technical Report TKN-04-16, A. Wolisz (Editor), Telecommunication Networks Group, Technical University Berlin, 43 p.

Gusenbauer, D., Isert, C. and Krösche, J. (2010): "Self-Contained Indoor Positioning on Off-The-Shelf Mobile Devices", Proceedings of the 2010 International Conference on Indoor Positioning and Indoor Navigation (IPIN), September 15-17, 2010 Campus Science City, ETH Zurich, Switzerland.

Gustafsson F. and Gunnarsson, F. (2005): "Mobile Positioning Using Wireless Networks," IEEE Signal Processing Magazine, vol. 22, no. 4, pp. 41-53.

Habbecke, M. and Kobbelt, L. (2008): "Laser Brush: a Flexible Device for 3D Reconstruction of Indoor Scenes", Symposium on Solid and Physical Modeling vol. 2008, pp. 231-239.

Hagisonic (2008): "User's Guide Localization System StarGazerTM for Intelligent Robots", http://www.hagisonic.com/, last accessed 17. March 2010.

Hansen, R., Wind, R., Jensen, C. and Thomsen, B. (2010): "Algorithmic Strategies for Adapting to Environmental Changes in 802.11 Location Fingerprinting", Proceedings of the 2010 International Conference on Indoor Positioning and Indoor Navigation (IPIN), September 15-17, 2010 Campus Science City, ETH Zurich, Switzerland.

Hashemi, H. (1993): "The Indoor Radio Propagation Channel," Proceedings of IEEE, vol. 81, no. 7, pp. 943-968.

Hauschildt, D. and Kirchhof, N. (2010): "Advances in Thermal Infrared Localization: Challenges and Solutions", Proceedings of the 2010 International Conference on Indoor Positioning and Indoor Navigation (IPIN), September 15–17, 2010 Campus Science City, ETH Zurich, Switzerland.

Hausmair, K., Witrisal, K., Meissner, P., Steiner, C. and Kail, G. (2010): "SAGE Algorithm for UWB Channel Parameter Estimation", Proceedings of the 10th COST 2100 Management Committee Meeting.

Haverinen, J. and Kemppainen, A. (2009): "Global Indoor Self-Localization Based on the Ambient Magnetic Field", Journal of Robotics and Autonomous Systems, vol. 57, no. 10, pp. 1028–1035.

Hazas, M. and Hopper, A. (2006): "Broadband Ultrasonic Location Systems for Improved Indoor Positioning", IEEE Transactions on Mobile Computing, vol. 5, no. 5, pp. 536–547.

Hein, G., Paonni, M., Kropp, V. and Teuber, A. (2008): "GNSS Indoors. Fighting the Fading – Part 1", Inside GNSS - Engineering Solutions for the Global Navigation Satellite System Community, vol. 3, no. 2, March/April 2008, pp. 43–52.

Herrmann, R., Sachs, J. and Bonitz, F. (2010): "On Benefits and Challenges of Person Localization Using Ultra-Wideband Sensors", Proceedings of the 2010 International Conference on Indoor Positioning and Indoor Navigation (IPIN), September 15–17, 2010 Campus Science City, ETH Zurich, Switzerland.

Hexamite (2011): http://www.hexamite.com/, last accessed 4. October 2011.

Hightower, J. and Borriello, G. (2001): "Location Systems for Ubiquitous Computing", Computer, IEEE Computer Society Press, vol. 34, no. 8, August 2001, pp. 57–66.

Hile, H. and Borriello, G. (2008): "Positioning and Orientierung in Indoor Environments Using Camera Phones", IEEE Computer Graphics and Applications, July/August 2008, pp. 32–39.

Ido, J., Shimizu, Y., Matsumoto, Y. and Ogasawara, T. (2009): "Indoor Navigation for a Humanoid Robot Using a View Sequence", The International Journal of Robotics Research, vol. 28, no. 2, pp. 315–325.

Ingensand, H., Bitzi, P. (2001): "Technologien der GSM-Positionierungsverfahren", AVN – Allgemeine Vermessungs-Nachrichten, vol. 2001, no. 8-9, pp. 286-294.

InfraSurvey (2011): http://www.infrasurvey.ch/, last accessed 12. June 2011.

JCGM 200:2008 (2008): "International Vocabulary of Metrology - Basic and General Concepts and Associated Terms", 3rd Edition, Joint Committee for Guides in Metrology, 104 p.

Jekeli, C. (2001): "Inertial Navigation Systems with Geodetic Applications", Walter de Gruyter, New York, NY, 352 p.

Joseph, A. (2010): "GNSS Solutions: Measuring GNSS Signal Strength", Inside GNSS - Engineering Solutions for the Global Navigation Satellite System Community, vol. 5, no. 8, pp. 20–25, Nov/Dec 2010.

Jiménez, A.R., Prieto, J.C., Ealo, J.L., Guevara J. and Seco, F. (2009): "A Computerized System to Determine the Provenance of Finds in Archaeological Sites Using Acoustic Signals", Journal of Archaeological Science, vol. 36, no. 10, pp. 2415–2426.

Jiménez, A.R., Seco, F., Prieto, J.C. and Guevara, J. (2010): "Pedestrian Indoor Navigation by Aiding a Foot-Mounted IMU with RFID Signal Strength Measurements", Proceedings of the 2010 International Conference on Indoor Positioning and Indoor Navigation (IPIN), September 15–17, 2010 Campus Science City, ETH Zurich, Switzerland.

Kee, C., Yun, D., Jun, H., Parkinson, B., Pullen, S. and Lagenstein, T. (2001): "Centimeter-Accuracy Indoor Navigation Using GPS-like Pseudolites," vol. 12, no. 11. Advanstar Communications Inc., pp. 14–23.

Kee, C., Jun, H. and Yun, D. (2003): "Indoor Navigation System Using Asynchronous Pseudolites", Journal of Navigation, vol. 56, pp. 443–455.

Kemppi, P., Rautiainen, T., Ranki, V., Belloni, F. and Pajunen, J. (2010): "Hybrid Positioning System Combining Angle-Based Localization, Pedestrian Dead Reckoning and Map Filtering", Proceedings of the 2010 International Conference on Indoor Positioning and Indoor Navigation (IPIN), September 15–17, 2010 Campus Science City, ETH Zurich, Switzerland.

Keßler, C., Ascher, C. and Trommer, G. (2011): "Multi-Sensor Indoor Navigation System with Vision- and Laser-Based Localisation and Mapping Capabilities", European Journal of Navigation, vol. 9, no. 3, pp. 4–11.

Khoshelham, K. (2010): "Automated Localization of a Laser Scanner in Indoor Environments Using Planar Objects", Proceedings of the 2010 International Conference on Indoor Positioning and Indoor Navigation (IPIN), September 15–17, 2010 Campus Science City, ETH Zurich, Switzerland.

Kiers, M., Krajnc, E., Dornhofer, M. and Bischof, W. (2011): "Evaluation and Improvements of an RFID Based Indoor Navigation System for Visually Impaired and Blind People", Proceedings of the 2011 International Conference on Indoor Positioning and Indoor Navigation (IPIN), September 21–23, 2011 in Guimarães, Portugal.

Kim, J. and Jun, H.S. (2008): "Vision-Based Location Positioning using Augmented Reality for Indoor Navigation", IEEE Transaction on Consumer Electronics, vol. 54, no. 3, pp. 954–962.

Kimaldi (2011): http://www.kimaldi.com/, last accessed 19. October 2010.

King, T., Kopf, S., Haenselmann, T., Lubberger, C. and Effelsberg, W. (2006): "Compass: A Probabilistic Indoor Positioning System Based on 802.11 and Digital Compasses", Proceedings of the First ACM International Workshop on Wireless Network Testbeds, Experimental Evaluation and Characterization (WiNTECH), Los Angeles, CA, USA, September 2006.

References

Kirschner, H. and Stempfhuber, W. (2008): "The Kinematic Potential of Modern Tracking Total Stations - A State of the Art Report on the Leica TPS1200+", in: Stempfhuber, W. and Ingensand, H. [Eds.], Proceedings of the 1st International Conference on Machine Control & Guidance, ETH Zurich, pp. 51–60.

Kitanov, A., Biševac, S. and Petrović, I. (2007): "Mobile Robot Self-Localization in Complex Indoor Environments Using Monocular Vision and 3D Model", Proceedings of the IEEE/ASME International Conference on Advanced Intelligent Mechatronics, Zürich, Switzerland.

Klingbeil, L., Romanovas, M., Schneider, P., Traechtler, M. and Manoli, Y. (2010): "A Modular and Mobile System for Indoor Localization", Proceedings of the 2010 International Conference on Indoor Positioning and Indoor Navigation (IPIN), September 15–17, 2010 Campus Science City, ETH Zurich, Switzerland.

Kjærgaard, M., Blunck, H., Godsk, T., Toftkjær, T, Christensen, D. and Grønbæk, K. (2010): "Indoor Positioning Using GPS Revisited", Pervasive Computing, Lecture Notes in Computer Science, vol. 6030, pp. 38–56.

Kocur, D., Rovňáková, J. and Švecová, M. (2009): "Through Wall Tracking of Moving Targets by M-Sequence UWB Radar", in: I. J. Rudas, J. Fodor, J, Kacprzyk [Eds.]: Computational Intelligence in Engineering, Springer, pp. 394–364.

Köhler, M., Patel, S., Summet, J., Stuntebeck E. and G. Abowed (2007): "TrackSense: Infrastructure Free Precise Indoor Positioning Using Projected Patterns", Pervasive Computing, LNCS, vol. 4480, pp. 334–350.

Kohoutek, T.K., Mautz, R. and Donaubauer, A. (2010): "Real-time Indoor Positioning Using Range Imaging Sensors", Proceedings of SPIE Photonics Europe, Real-Time Image and Video Processing, vol. 7724.

Kokeisl (2011): http://www.kokeisl.net/, last accessed 12. October 2011.

Kosch, O., Thiel, F., Ittermann, B. and Seifert, F. (2011): "Non-Contact Cardiac Gating with Ultra-Wideband Radar Sensors for High Field MRI", Proceedings of the 2011 Annual Meeting of the International Society for Magnetic Resonance in Medicine (ISMRM), no. 19, p. 1804.

Koski, L. Perälä, T. and Piché, R (2010): "Indoor Positioning Using WLAN Coverage Area Estimates", Proceedings of the 2010 International Conference on Indoor Positioning and Indoor Navigation (IPIN), September 15–17, 2010 Campus Science City, ETH Zurich, Switzerland.

Kranz, M., Fischer, C. and Schmidt, A. (2010): "A Comparative Study of DECT and WLAN Signals for Indoor Localization", Proceedings of the IEEE International Conference on Pervasive Computing and Communications (PerCom 2010), pp. 235–243.

Kröll, H. and Steiner, C. (2010): "Indoor Ultra-Wideband Location Fingerprinting", Proceedings of the 2010 International Conference on Indoor Positioning and Indoor Navigation (IPIN), September 15–17, 2010 Campus Science City, ETH Zurich, Switzerland.

Lachapelle, G. (2004): "GNSS Indoor Location Technologies", Journal of Global Positioning Systems vol. 3, no. 1–2, pp. 2–11.

Laitinen, H. (2004): "WLAN Location Methods," Graduate Course Slides, University of Finland, online available at http://www.comlab.hut.fi/opetus/333/2004slides/topic33.pdf, last accessed 3. December 2011.

Lambda:4 (2011): http://www.lambda4.com/EN/, last accessed 11. October 2011.

Larrañaga, J., Muguira, L., Lopez-Garde, J.M. and Vazquez, J.I. (2010): "An Environment Adaptive ZigBee-Based Indoor Positioning Algorithm", Proceedings of the 2010 International Conference on Indoor Positioning and Indoor Navigation (IPIN), September 15–17, 2010 Campus Science City, ETH Zurich, Switzerland.

Lee, S. and Song, J.B. (2007): "Mobile Robot Localization Using Infrared Light Reflecting Landmarks", Proceedings of the International Conference on Control, Automation and Systems (ICCAS'07), pp. 674–677.

Leica Geosystems (2011): http://www.leica-geosystems.ch/, last accessed 19. December 2011.

Liu, X., Makino, H., Kobyashi, S. and Maeda, Y. (2006): "An Indoor Guidance System for the Blind using Fluorescent Lights - Relationship between Receiving Signal and Walking Speed", Proceedings of the 28th IEEE EMBS Conference, pp. 5960–5963.

Liu, H., Darabi, H., Banerjee, P. and Liu, J. (2007): "Survey of Wireless Indoor Positioning Techniques and Systems", IEEE Transactions on Systems, Man, and Cybernetics – Part C: Applications and Reviews, vol. 37, no. 6, pp. 1067–1080.

Liu, T., Carlberg, M., Chen, G., Chen, J., Kua, J. and Zakhor, A. (2010): "Indoor Localization and Visualization Using a Human-Operated Backpack System", Proceedings of the 2010 International Conference on Indoor Positioning and Indoor Navigation (IPIN), September 15–17, 2010 Campus Science City, ETH Zurich, Switzerland, pp. 890–899.

Liu, W., Hu, C., He, Q., Meng, M. and Liu, L. (2010): "A Hybrid Localization System Based on Optics and Magnetics", Proceedings of the International Conference on Robotics and Biomimetics, Tianjin, China, pp. 1165–1169.

Locata (2011): http://www.locatacorp.com/, last accessed 29. October 2011.

Loctronix (2011): http://www.loctronix.com/, last accessed 11. December 2011.

Lukianto, C., Hönniger, C. and Sternberg, H. (2010): "Pedestrian Smartphone-Based Indoor Navigation Using Ultra Portable Sensory Equipment", Proceedings of the 2010 International Conference on Indoor Positioning and Indoor Navigation (IPIN), September 15–17, 2010 Campus Science City, ETH Zurich, Switzerland.

Mandal, A., Lopes, C.V., Givargis, T., Haghighat, A., Jurdak, R. and Baldi, P. (2005): "Beep: 3D Indoor Positioning Using Audible Sound", Proceedings of the Consumer Communications and Networking Conference (CCNC 2005), IEEE Xplore.

Mathiassen, K., Hanssen, L. and Hallingstad, O. (2010): "A Low Cost Navigation Unit for Positioning of Personnel After Loss of GPS Position", Proceedings of the 2010 International Conference on Indoor Positioning and Indoor Navigation (IPIN), September 15–17, 2010 Campus Science City, ETH Zurich, Switzerland.

Mautz, R. (2002): "Solving Nonlinear Adjustment Problems by Global Optimization", Bollettino di Geodesia e Scienze Affini, vol. 61, no. 2, pp. 123 134.

Mautz, R. (2005): "Service Requirements Document (SRD)" for the EPSRC funded project "intelligent Pervasive LOcation Tracking (iPLOT)", Final Report, unpublished document, Imperial College London, UK, 50 p.

Mautz, R. and Ochieng, W.Y. (2007): "A Robust Indoor Positioning and Auto-Localisation Algorithm", Journal of Global Positioning Systems, vol. 6, no. 1, pp. 38–46.

Mautz, R., Ochieng, W.Y., Brodin, G., Kemp, A. (2007a): "3D Wireless Network Localization from Inconsistent Distance Observations," Ad Hoc & Sensor Wireless Networks, vol. 3, no. 2–3, pp. 141–170.

Mautz, R. (2009): The Challenges of Indoor Environments and Specification on some Alternative Positioning Systems, Proceedings of the 6th Workshop on Positioning, Navigation and Communication 2009 (WPNC'09), IEEE Xplore, pp. 29–36.

Mautz, R. and Tilch, S. (2011): "Optical Indoor Positioning Systems", Proceedings of the 2011 International Conference on Indoor Positioning and Indoor Navigation (IPIN), September 21–23, 2011 in Guimarães, Portugal.

Maye, O., Schaeffner, J. and Maaser, M. (2006): "An Optical Indoor Positioning System for the Mass Market", Proceedings of the 3rd Workshop on Positioning, Navigation and Communication (WPNC'06), 16 March 2006 in Hanover, Germany, IEEE Xplore, pp. 111–115.

Mazuelas, S., Bahillo, A., Lorenzo, R.M., Fernandez, P., Lago, F.A., Garcia, E., Blas, J. and Abril, J. (2009): "Robust Indoor Positioning Provided by Real-Time RSSI Values in Unmodified WLAN Networks," IEEE Journal of Selected Topics in Signal Processing, vol. 3, no. 5, pp. 821–831.

Microsoft Kinect (2011): http://www.xbox.com/de-DE/kinect/ last accessed 11. October 2011.

Moghtadaiee, V., Dempster, A. and Lim, S. (2011): "Indoor Localization Using FM Radio Signals: A Fingerprinting Approach", Proceedings of the 2011 International Conference on Indoor Positioning and Indoor Navigation (IPIN), September 21–23, 2011 in Guimarães, Portugal.

Molisch, A. (2009): "Ultrawideband Propagation Channels", Proceedings of IEEE, Special Issue on UWB, vol. 97, pp. 353–371.

Mulloni, A., Wgner, D., Schmalstieg, D. and Barakonyi, I. (2009): "Indoor Positioning and Navigation with Camera Phones", Pervasive Computing, IEEE, vol. 8, pp. 22–31.

Muffert, M., Siegemund, J. and Förstner, W. (2010): "The Estimation of Spatial Positions by Using an Omnidirectional Camera System", Proceedings of the 2nd International Conference on Machine Control & Guidance, pp. 95–104.

Muthukrishnan, K., Koprinkov, G., Meratnia, N. and Lijding, M. (2006): "Using Time-of-Flight for WLAN Localization: Feasibility Study", Centre for Telematics and Information Technology (CTIT), University of Twente, Technical Report, no. TR-CTIT-06-28.

My-Bodyguard (2011): www.my-bodyguard.eu, last accessed 21. October 2011.

Nanotron Technologies (2011): www.nanotron.com, last accessed 22. October 2011.

NaviFloor (2011): www.future-shape.com/en/navifloor.html, last accessed 22. October 2011.

Nikon Metrology (2011): http://www.nikonmetrology.com/, last accessed 1. March 2011.

Nilsson, J.O., Skog, I. and Händel, P. (2010): "Performance Characterisation of Foot-Mounted ZUPT-Aided INSs and Other Related Systems", Proceedings of the 2010 International Conference on Indoor Positioning and Indoor Navigation (IPIN), September 15–17, 2010 Campus Science City, ETH Zurich, Switzerland.

Nishikata, H., Makino, H., Nishimori, K., Kaneda, T., Liu, X., Kobayashi, M. and Wakatsuki, D. (2011): "Basic Research of Indoor Positioning Method Using Visible Light Communication and Dead Reckoning", Proceedings of the 2011 International Conference on Indoor Positioning and Indoor Navigation (IPIN), September 21–23, 2011 in Guimarães, Portugal.

Niwa, H., Kodaka, K., Sakamoto, Y., Otake, M., Kawaguchi, S., Fujii, K., Kanemori, Y. and Sugano, S. (2008): "GPS-Based Indoor Positioning System with Multi-Channel Pseudolite", Proceedings of the IEEE International Conference of Robotics and Automation (ICRA), pp. 905–910.

OKI (2011): http://www.oki.com/, last accessed 16. November 2011.

Papliatseyeu, A., Kotilainen, N., Mayora, O. and Osmani, V. (2009): "FINDR: Low-Cost Indoor Positioning Using FM Radio", MobileWireless Middleware, Operating Systems, and Applications (2009), pp. 15–26.

Patel, S., Stuntebeck, E. and Robertson, T. (2009): "PL-Tags: Detecting Batteryless Tags through the Power Lines in a Building", Proceedings of the International Conference on Pervasive Computing (Pervasive '09), pp. 257–273.

Park, J., Charrow, B., Curtis, D., Battat, J., Minkov, E., Hicks, J., Teller, S. and Ledlie, J. (2010): "Growing an Organic Indoor Location System," Proceedings of International Conference on Mobile Systems, Applications, and Services, pp. 271–284.

Parodi, B. B., Lenz, H., Szabo, A., Wang, H., Horn, J., Bamberger, J. and Obradovic J. (2006): "Initialization and Online-Learning of RSS Maps for Indoor/Campus Localization," Proceedings of Position Location and Navigation Symposium (PLANS 06), IEEE/ION, Myrtle Beach, South Carolina pp. 24–27.

Pena, D., Feick, R., Hristov, H. and Grote, W. (2003): "Measurement and Modelling of Propagation Losses in Brick and Concrete Walls for the 900-MHz Band," IEEE Transactions on Antennas and Propagation, vol. 51, no. 1, pp. 31–39.

Peng, J., Zhu, M. and Zhang, K. (2011): "New Algorithms Based on Sigma Point Kalman Filter Technique for Multi-sensor Integrated RFID Indoor/Outdoor Positioning", Proceedings of the 2011 International Conference on Indoor Positioning and Indoor Navigation (IPIN), September 21–23, 2011 in Guimarães, Portugal.

Pereira, F., Theis, C., Moreira, A. and Ricardo, M. (2011): "Evaluating Location Fingerprinting Methods for Underground GSM Networks deployed over Leaky Feeder", Proceedings of the 2011 International Conference on Indoor Positioning and Indoor Navigation (IPIN), September 21–23, 2011 in Guimarães, Portugal.

Pietrzyk, M. and von der Grün, T. (2010): "Experimental Validation of a TOA UWB Ranging Platform with the Energy Detection Receiver", Proceedings of the 2010 International Conference on Indoor Positioning and Indoor Navigation (IPIN), September 15–17, 2010 Campus Science City, ETH Zurich, Switzerland.

Polhemus (2011): http://www.polhemus.com/, last accessed 30. May 2011.

Popescu, V., Sacks, E. and Bhamutov, G. (2004): "Interactive Modeling from Dense Color and Sparse Depth", Proceedings of Conference 3D Imaging, Modeling, Processing, Visualization and Transmission (3DPVT), pp. 430–437.

Popescu, V., Bahmutov, G., Mudure, M. and Sacks, E. (2006): "The Modelcamera", Graphical Models, vol. 68, pp. 385–401.

Popleteev, A. (2011): "Indoor Positioning Using FM Radio Signals", PhD Dissertation at the University of Trento, School in Information and Communication Technologies, 160 p.

Povalač, A. and Šebesta, J. (2010): "Phase of Arrival Ranging Method for UHF RFID Tags Using Instantaneous Frequency Measurement," Proceedings of the 20th International Conference on Applied Electromagnetics and Communications (ICECom 2010), vol. 1, pp. 1–4.

Pressl, B. and Wieser, M. (2006): "A Computer-Based Navigation System Tailored to the Needs of Blind People", Lecture Notes in Computer Science, vol. 4061, pp. 1280–1286.

Priyantha, N.B. (2005): "The Cricket Indoor Location System", PhD Thesis, Massachusetts Institute of Technology, 199p.

Quddus, M., Ochieng, W. and Noland, R. (2007): "Current Map-Matching Algorithms for Transport Applications: State of the Art and Future Research Directions" Transportation Research Part C, vol. 15, pp. 312–328.

Rabinowitz, M. and Spilker, J. (2005): "A New Positioning System Using Television Synchronization Signals", IEEE Transactions on Broadcasting, vol. 51, no. 1, pp. 51-61.

Rantakokko, J., Händel, P., Fredholm, M. and Marsten-Eklöf, F. (2010): "User Requirements for Localization and Tracking Technology: A Survey of Mission-specific Needs and Constraints", Proceedings of the 2010 International Conference on Indoor

References

Positioning and Indoor Navigation (IPIN), September 15–17, 2010 Campus Science City, ETH Zurich, Switzerland.

Reddy, H. and Chandra, G.: (2007): "An Improved Time-of-Arrival Estimation for WLAN-Based Local Positioning", Proceedings of the 2nd International Conference on Communication Systems Software and Middleware (COMSWARE), January 2007, pp. 1–5.

Reijiniers, J. and Peremans, H. (2007): "Biomimetic Sonar System Performing Spectrum-Based Localization", IEEE Transactions on Robotics, vol. 23, no. 6, pp. 1151–1159.

Reimann, R. (2011): "Locating and Distance Measurement by High Frequency Radio Waves", Proceedings of the 2011 International Conference on Indoor Positioning and Indoor Navigation (IPIN), September 21–23, 2011 in Guimarães, Portugal.

Renaudin, V., Yalak, O. and Tomé, P. (2007): "Hybridization of MEMS and Assisted GPS for Pedestrian Navigation," Inside GNSS, vol. 2, no. 1, pp. 34–42.

Richardson, B., Leydon, K., Fernstrom, M. and Paradiso, J. (2004): "Z-Tiles: Building Blocks for Modular, Pressure-Sensing Floorspaces", Proceedings of the ACM Conference on Human Factors and Computing Systems (CHI 2004), pp. 1529–1532.

Rimminen, H., Lndström, J., and Sepponen, R. (2009): "Positioning Accuracy and Multi-Target Separation with a Human Tracking System Using Near Field Imaging", International Journal On Smart Sensing and Intelligent Systems, vol. 2, no. 1, pp. 156–175.

Rimminen, H., Lndström, J., Linnavuo, M. and Sepponen, R. (2010): "Detection of Falls Among the Elderly by a Floor Sensor Using the Electric Near Field", IEEE Trans Transactions on Information Technology in Biomedicine vol. 14, no. 6, pp. 1475–1476.

Rizos, C., Roberts, G., Barnes, J. and Gambale, N. (2010): "Locata: A new High Accuracy Indoor Positioning System", Proceedings of the 2010 International Conference on Indoor Positioning and Indoor Navigation (IPIN), September 15–17, 2010 Campus Science City, ETH Zurich, Switzerland.

Robert, C., Tomé, P., Botteron, C., Farine, P.A., C., Merz, R. and Blatter, A. (2010): "Low Power ASIC Transmitter for UWB-IR Radio Communication and Positioning", Proceedings of the 2010 International Conference on Indoor Positioning and Indoor Navigation (IPIN), September 15–17, 2010 Campus Science City, ETH Zurich, Switzerland.

Röhr, S. and Gulden, P. (2009): "Hochpräzise Kranortung mit Millimeterwellen", 17. Kranfachtagung at the Technical University of Dresden, 16 p.

Salo, J., Vuokko, L., El-Sallabi, H. M. and Vainikainen, P. (2007): "An Additive Model as a Physical Basis for Shadow Fading," IEEE Transactions on Vehicular Technology, vol. 56, no. 1, pp. 13–26.

Sato, T., Nakamura, S., Terabayashi, K., Sugimoto, M. and Hashizume. H. (2011): "Design and Implementation of a Robust and Real-time Ultrasonic Motion-capture System",

Proceedings of the 2011 International Conference on Indoor Positioning and Indoor Navigation (IPIN), September 21–23, 2011 in Guimarães, Portugal.

Schantz, H.G., Christian, W. and John, U. (2011): "Characterization of Error in a NFER Real-Time Location System", Proceedings of Radio and Wireless Symposium, IEEE, 16–20 January 2011.

Schlaile, C., Meister, O., Frietsch, N., Kessler, C., Wendel, J. and Trommer, G. F. (2009): "Using Natural Features for Vision Based Navigation of an Indoor-VTOL MAV", Aerospace Science and Technology, vol. 13, no. 7, pp. 349–357.

Schmid, J., Völker, M., Gädeke, T., Weber, P., Stork, W. and Müller-Glaser, K. (2010): "An Approach to Infrastructure-Independent Person Localization with an IEEE 802.15.4 WSN", Proceedings of the 2010 International Conference on Indoor Positioning and Indoor Navigation (IPIN), September 15–17, 2010 Campus Science City, ETH Zurich, Switzerland.

Schmitt, R., Nisch, S., Schönberg, A., Demeester, F. and Renders, S. (2010): "Performance Evaluation of iGPS for Industrial Applications", Proceedings of the 2010 International Conference on Indoor Positioning and Indoor Navigation (IPIN), September 15–17, 2010 Campus Science City, ETH Zurich, Switzerland.

Schneider, O. (2010): "Requirements for Positioning and Navigation in Underground Constructions", Proceedings of the 2010 International Conference on Indoor Positioning and Indoor Navigation (IPIN), September 15–17, 2010 Campus Science City, ETH Zurich, Switzerland.

Schwaighofer, A., Grigoras, M., Tresp, V. and Hoffmann, C. (2003): "GPPS: A Gaussian Process Positioning System for Cellular Networks", in Advances in Neural Information Processing Systems, 2003.

Schweinzer, H. and Syafrudin S. (2010): "LOSNUS: An Ultrasonic System Enabling High Accuracy and Secure TDoA Locating of Numerous Devices", Proceedings of the 2010 International Conference on Indoor Positioning and Indoor Navigation (IPIN), September 15–17, 2010 Campus Science City, ETH Zurich, Switzerland.

Seco, F., Jiménez, A. R., Prieto, C., Roa J. and Koutsou, K. (2009): "A Survey of Mathematical Methods for Indoor Localization", 6th IEEE International Symposium on Intelligent Signal Processing (WISP 2009), Budapest, Hungary, pp. 9–14.

Seco, F., Plagemann, C., Jiménez, A. and Burgard, W. (2010): "Improving RFID-Based Indoor Positioning Accuracy Using Gaussian Processes", Proceedings of the 2010 International Conference on Indoor Positioning and Indoor Navigation (IPIN), September 15–17, 2010 Campus Science City, ETH Zurich, Switzerland.

Segura, M., Hashemi, H., Sisterna, C. and Mut, V. (2010): "Experimental Demonstration of Self-Localized Ultra Wideband Indoor Mobile Robot Navigation System", Proceedings of the 2010 International Conference on Indoor Positioning and Indoor Navigation (IPIN), September 15–17, 2010 Campus Science City, ETH Zurich, Switzerland.

Seitz, J., Vaupel, T., Meyer, S., Gutiérrez Boronat, J. and Thielecke, J. (2010): "A Hidden Markov Model for Pedestrian Navigation," Proceedings of the 7th Workshop on Positioning, Navigation and Communication (WPNC' 10), Dresden, Germany.

SensFloor (2011): http://www.future-shape.com/en/, last accessed 22. October 2011.

Serant, D., Julien, O., Ries, L., Thevenon, P. and Dervin, M. (2010): "The Digital TV Case", Inside GNSS - Engineering Solutions for the Global Navigation Satellite System Community, vol. 6, no. 6, pp. 54–62, Nov/Dec 2011.

Shang, Y., Ruml, W., Zhang, Y. and Fromherz, M. (2004): "Localization from Connectivity in Sensor Networks", IEEE Transactions on Parallel and Distributed Systems, vol. 15, no. 11, pp. 961–974.

Shen, J., Oda, Y. (2010): "Direction Estimation for Cellular Enhanced Cell-ID Positioning Using Multiple Sector Observations", Proceedings of the 2010 International Conference on Indoor Positioning and Indoor Navigation (IPIN), September 15–17, 2010 Campus Science City, ETH Zurich, Switzerland.

Shin, S., Kim, H.W., Park, C.G. and Yoo, Y.M. (2009): "Sit-Down and Stand-Up Awareness Algorithm for the Pedestrian Dead Reckoning," Proceedings of the European Navigation Conference on Global Navigation Satellite Systems (ENC-GNSS'09).

Sjö, K., López, D., Paul, C., Jensfelt, P. and Kragic D. (2009): "Object Search and Localization for an Indoor Mobile Robot", Journal of Computing and Information Technology, vol. 17, no. 1, pp. 67–80.

Skog, I., Nilsson, J. O. and Händel, P. (2010): "Evaluation of Zero-Velocity Detectors for Foot-Mounted Inertial Navigation Systems", Proceedings of the 2010 International Conference on Indoor Positioning and Indoor Navigation (IPIN), September 15–17, 2010 Campus Science City, ETH Zurich, Switzerland.

Skyhook (2011): www.skyhookwireless.com, last accessed 15. November 2011.

Sky-Trax Inc. (2011): http://www.sky-trax.com/, last accessed 22. October 2011.

Solcon (2011): http://www.solcon-systemtechnik.de/, last accessed 22. October 2011.

SoLoc (2009): "Signals of Opportunity Tracking Device", Project Report, San Diego State University, 30 p.

Soloviev, A. and Dickman, J. (2010): "Deeply Integrated GPS for Indoor Navigation", Proceedings of the 2010 International Conference on Indoor Positioning and Indoor Navigation (IPIN), September 15–17, 2010 Campus Science City, ETH Zurich, Switzerland.

Soloviev, A. and Venable, D. (2010): "When GNSS Goes Blind - Integrating Vision Measurements for Navigation in Signal-Challenged Environments", GNSS Inside, October 2010, pp. 18–29.

Song, S., Hu, C., Mao, L., Yang, W. and Meng, M. (2009): "Real Time Algorithm for Magnet's Localization in Capsule Endoscope", Proceedings of the International Conference on Automation and Logistics (ICAL'09), Shenyang, China, pp. 2030–2035.

Sonitor (2011): http://www.sonitor.com/, last accessed October 2011.

Steinhage, A. and Lauterbach, C. (2007) "SensFloor – Ein großflächiges Sensorsystem für Ambient-Assisted-Living Anwendungen," Proceedings of the MikroSystemTechnik Kongress, 2007, pp. 1091- 1094.

Stelzer, A., Pourvoyeur, K. and Fischer, A. (2004): "Concept and Application of LPM – A Novel 3-D Local Position Measurement System", IEEE Transactions on Microwave Theory and Techniques, vol. 52, no. 12, pp. 2664–2669.

Stoica, L., Rabbachin, A. and Oppermann, I. (2006): „A Low-Complexity Noncoherent IR-UWB Transceiver Architecture with TOA Estimation", IEEE Transactions on Microwave Theory and Techniques, vol. 54, no. 4, April 2006, pp. 1637–1646.

Stone, W. (1997): "Electromagnetic Signal Attenuation in Construction Materials", NIST Report 6055, National Institute of Standards, Gaithersburg, Maryland, no. 185, 8 p.

Sun, Z., Mao, X., Tian, W. and Zhang, X. (2009): "Activity Classification and Dead Reckoning for Pedestrian Navigation with Wearable Sensors," Measurement Science and Technology, vol. 20, no. 1, 2009, pp. 1–10.

Stuntebeck, E., Patel, S., Robertson, T., Reynolds, M. and Abowd, G. (2008): "Wideband PowerLine Positioning for Indoor Localization", International Conference on Ubiquitous Computing (UbiComp '08), pp. 94–103.

Symeo (2011): http://www.symeo.com/, last accessed 26. February 2011.

Tadakamadla, S. (2006): "Indoor Local Positioning System for Zigbee Based on RSSI", M.Sc. Thesis Report, Mid Sweden University, 50 p.

Tappero, F. (2009): "Low-Cost Optical-Based Indoor Tracking Device for Detection and Mitigation of NLOS Effects", Procedia Chemistry, Elsevier, vol. 1, no.1, pp. 497–500.

Tappero, F., Merminod, B. and Ciurana, M. (2010): "IEEE 802.11 Ranging and Multi-lateration for Software-Defined Positioning Receiver", Proceedings of the 2010 International Conference on Indoor Positioning and Indoor Navigation (IPIN), September 15–17, 2010 Campus Science City, ETH Zurich, Switzerland.

Teuber A. and Eissfeller, B. (2006): "WLAN Indoor Positioning Based on Euclidean Distances and Fuzzy Logic," Proceedings of the 3rd Workshop on Positioning, Navigation and Communication (WPNC'06), Hannover, Germany.

Tilch, S. and Mautz, R. (2010): "Development of a New Laser-Based, Optical Indoor Positioning System", Proceedings of the ISPRS Commission V Mid-Term Symposium Close Range Image Measurement Techniques, no. 98, pp. 575–580.

Trimble (2011): http://www.trimble.com/mining/Terralite-XPS-Solutions/, last accessed 31. October 2011.

Trucco, E. and Plakas, K. (2006): "Video Tracking: A Concise Survey", IEEE Journal of Oceanic Engineering, vol. 31, no. 2, pp. 520–529.

Ubisense (2011): http://www.ubisense.net/en/, last accessed 4. February 2011.

Uchitomi, N., Inada, A., Fujimoto, M., Wada, T., Mutsuura, K. and Okada, H. (2010): "Accurate Indoor Position Estimation by Swift-Communication Range Recognition (S-

CRR) Method in Passive RFID systems", Proceedings of the 2010 International Conference on Indoor Positioning and Indoor Navigation (IPIN), September 15–17, 2010 Campus Science City, ETH Zurich, Switzerland.

Valtonen, M., Mäentausta J. and Vanhala J. (2009): "TileTrack: Capacitive Human Tracking Using Floor Tiles", Proceedings of the Seventh Annual IEEE International Conference on Pervasive Computing and Communications, 2009, pp. 31–40.

Varshavsky, A., de Lara, E., LaMarca, A., Hightower, J. and Otsason, V., (2007): "GSM Indoor Localization", Pervasive and Mobile Computing Journal (PMC), Elsevier, vol. 3, no. 6, pp. 698–720.

Vaupel, T., Seitz, J., Kiefer, F., Haimerl, S. and Thielecke, J. (2010): "Wi-Fi Positioning: System Considerations and Device Calibration", Proceedings of the 2010 International Conference on Indoor Positioning and Indoor Navigation (IPIN), September 15–17, 2010 Campus Science City, ETH Zurich, Switzerland.

Vervisch-Picois, A., Samama, N. (2009): "Indoor Carrier Phase Measurements through GNSS Transmitters Theory and First Experimental Results", Proceedings of the Congress of the International Association of Institutes of Navigation (IAIN2009), Stockholm, Sweden.

Wagner, J., Isert, C., Purschwitz, A. and Kistner, A. (2010): "Improved Vehicle Positioning for Indoor Navigation in Parking Garages Through Commercially Available Maps", Proceedings of the 2010 International Conference on Indoor Positioning and Indoor Navigation (IPIN), September 15–17, 2010 Campus Science City, ETH Zurich, Switzerland.

Walder, U. and Bernoulli, T. (2010): "Context-Adaptive Algorithms to Improve Indoor Positioning with Inertial Sensors", Proceedings of the 2010 International Conference on Indoor Positioning and Indoor Navigation (IPIN), September 15–17, 2010 Campus Science City, ETH Zurich, Switzerland.

Wan, E. and Paul, A. (2010): "A Tag-free Solution to Unobtrusive Indoor Tracking Using Wall-mounted Ultrasonic Transducers", Proceedings of the 2010 International Conference on Indoor Positioning and Indoor Navigation (IPIN), September 15–17, 2010 Campus Science City, ETH Zurich, Switzerland.

Wan, S. and Foxlin, E. (2010): "Improved Pedestrian Navigation Based on Drift-Reduced MEMS IMU Chip", Proceedings of ION 2010 International Technical Meeting, January 25–27, 2010, San Diego, California, pp. 365–374.

Wang, S., Waadt, A., Burnic, A., Dong Xu, Kocks, C., Bruck, G.H. and Jung, P. (2010): "System Implementation Study on RSSI Based Positioning in UWB Networks", 7th International Symposium on Wireless Communication Systems (ISWCS), pp. 36–40.

Want, R., Hopper, A., Falcão, V. and Gibbons, J. (1992): "The Active Badge Location System", ACM Transactions on Information Systems, vol. 10, no. 1, pp. 91–102.

Ward, A., Jones, A. and Hopper, A. (1997): "A New Location Technique for the Active Office", Personal Communications, IEEE, vol. 4, no. 5, October 1997, pp. 42–47.

References

Weber, M., Birkel, U., Collmann, R. and Engelbrecht, J. (2011): "Wireless Indoor Positioning: Localization Improvements with a Leaky Coaxial Cable", Proceedings of the 2011 International Conference on Indoor Positioning and Indoor Navigation (IPIN), September 21–23, 2011 in Guimarães, Portugal.

Wieser, A. (2006): "High-Sensitivity GNSS: The Trade-Off between Availability and Accuracy", in: Kahmen H. and Chrzanowski A. [Eds.] Proceedings of the 3rd IAG / 12th FIG Symposium, Baden, Austria.

Wirola, L., Laine, T. and Syrjärinne J. (2010): "Mass Market Considerations for Indoor Positioning and Navigation", Proceedings of the 2010 International Conference on Indoor Positioning and Indoor Navigation (IPIN), September 15–17, 2010 Campus Science City, ETH Zurich, Switzerland.

Wong, C., Klukas, R. and Messier, G. (2008): "Using WLAN Infrastructure for Angle-of-Arrival Indoor User Location", Proceedings of the IEEE Vehicular Technology Conference, VTC 2008-Fall, pp. 1–5, Calgary, BC, September 2008.

Xiang, Z., Song, S., Chen, J., Wang, H., Huang, J. and Gao, X. (2004): "A Wireless LAN-Based Indoor Positioning Technology", Journal of Research Development, vol. 48, no. 5/6, pp. 617–626.

Yick, J, Mukherjee, B. and Ghosal, D. (2008): "Wireless Sensor Network Survey", Computer Networks, vol. 52, no. 12, pp. 2292–2330.

Yokoo, K., Beauregard, S. and Schneider, M. (2009): "Indoor Relative Localization with Mobile Short-Range Radar", Proceedings of the Vehicular Technology Conference (VTC'09), pp. 1–5.

Zhang, J., Li, B., Dempster, A. and Rizos C. (2011): "Evaluation of High Sensitivity GPS Receivers", Coordinates – Positioning, Navigation and Beyond, vol. 7, no. 3, pp. 7–12.

Zebra Enterprise Solutions (2011): http://zes.zebra.com/ last accessed 25. February 2011.

Zetik, R., Shen, G. and Thomä, R. (2010): "Evaluation of Requirements for UWB Localization Systems in Home-Entertainment Applications", Proceedings of the 2010 International Conference on Indoor Positioning and Indoor Navigation (IPIN), September 15–17, 2010 Campus Science City, ETH Zurich, Switzerland.

ZONITH (2011): http://www.zonith.com/products/ips/, last accessed 21. October 2011.

i want morebooks!

Buy your books fast and straightforward online - at one of world's fastest growing online book stores! Environmentally sound due to Print-on-Demand technologies.

Buy your books online at
www.get-morebooks.com

Kaufen Sie Ihre Bücher schnell und unkompliziert online – auf einer der am schnellsten wachsenden Buchhandelsplattformen weltweit! Dank Print-On-Demand umwelt- und ressourcenschonend produziert.

Bücher schneller online kaufen
www.morebooks.de

VDM Verlagsservicegesellschaft mbH
Heinrich-Böcking-Str. 6-8   Telefon: +49 681 3720 174   info@vdm-vsg.de
D - 66121 Saarbrücken   Telefax: +49 681 3720 1749   www.vdm-vsg.de

Printed by Books on Demand GmbH, Norderstedt / Germany